PROTECTING OUR PLANET

LAND PRESERVATION

BY A. W. BUCKEY

Essential Library

An Imprint of Abdo Publishing
abdobooks.com

ABDOBOOKS.COM

Published by Abdo Publishing, a division of ABDO, PO Box 398166, Minneapolis, Minnesota 55439.

Printed in the United States of America, North Mankato, Minnesota.

052024
092024

THIS BOOK CONTAINS RECYCLED MATERIALS

Cover Photo: Reforestation Collection/Shutterstock Images
Interior Photos: Rob Hainer/Shutterstock Images, 4; Mark Kostich/Shutterstock Images, 7; Drew Angerer/Getty Images News/Getty Images, 9; Shutterstock Images, 10, 22, 26, 52, 66, 82, 101; Pierre Banoori/Shutterstock Images, 13; Patrick Pleul/dpa-Zentralbild/picture alliance/Getty Images, 14; Richard Whitcombe/Shutterstock Images, 18; Arnaud Finistre/AFP/Getty Images, 21; Matthias Breiter/Design Pics Editorial/Universal Images Group/Getty Images, 25; Historical Picture Archive/Corbis Historical/Getty Images, 30; Hulton Archive/Getty Images, 33; Martin Capek/Shutterstock Images, 35; Zhao Qin/iStockphoto, 38; Juancho Torres/Anadolu Agency/Getty Images, 42; Ververidis Vasilis/Shutterstock Images, 44; Rodrigo Buendia/AFP/Getty Images, 47; Manan Vatsyayana/AFP/Getty Images, 49; Trees4Goals, 51; Matt May/Alamy, 55; Zaruba Ondrej/Shutterstock Images, 56–57; I. Noyan Yilmaz/Shutterstock Images, 59; Chaideer Mahyuddin/AFP/Getty Images, 62; Mariusz Bugno/Shutterstock Images, 65; Eugen Haag/Shutterstock Images, 69; Ashley Cooper/Global Warming Images/Alamy, 70; Zhou Shengbin/Xinhua News Agency/Getty Images, 73; Travis Morisse/The Hutchinson News/AP Images, 77; Kris Grabiec/Shutterstock Images, 78; Bernhard Staehli/Shutterstock Images, 81; Ralph Lee Hopkins/Design Pics Editorial/Universal Images Group/Getty Images, 86; Juan Barreto/AFP/Getty Images, 89; Sebastien Burel/Shutterstock Images, 90; Kevin Sullivan/Digital First Media/Orange County Register/MediaNews Group/Getty Images, 94; Kevin Winter/Getty Images Entertainment/Getty Images, 98; David Jeffrey Ringer/Shutterstock Images, 100

Editor: Haley Williams
Series Designer: Cynthia Della-Rovere

Library of Congress Control Number: 2023949535

PUBLISHER'S CATALOGING-IN-PUBLICATION DATA
Names: Buckey, A. W., author.
Title: Land preservation / by A. W. Buckey
Description: Minneapolis, Minnesota: Abdo Publishing, 2025 | Series: Protecting our planet | Includes online resources and index.
Identifiers: ISBN 9781098293444 (lib. bdg.) | ISBN 9798384912712 (ebook)
Subjects: LCSH: Biosphere reserves--Juvenile literature. | Forest reserves--Juvenile literature. | Land trusts--Juvenile literature. | National parks and reserves--Juvenile literature. | Conservation of natural resources--Juvenile literature. | Open spaces--Government policy--Juvenile literature. | Environmental sciences--Juvenile literature.
Classification: DDC 333.951--dc23

CONTENTS

Bald cypress trees are commonly found in the swamps of the southeastern United States.

CHAPTER 1

SAVE THE SWAMP

It was early Sunday morning, and the sky was just beginning to streak pink with dawn light. Sara had been awake for an hour getting breakfast ready. Her older brother, Cyrus, was back in town. He was in the garage getting out the kayaks. Ever since they were kids, the siblings loved spending weekends taking boat trips down the bayou into Saulnier Swamps.

Sara lived on the coast of Louisiana in a town near New Orleans. Her house was a short walk from a piece of land that everyone called Saulnier Swamps, after the family that owned it. It was a twisting body of water, smaller than a river but bigger than a stream, lined with deep-rooted trees that clung to the muddy banks. Sara and Cyrus could hear the chirp of the birds as they dragged their kayaks down a gravel path to the water. No matter how many times Sara traveled into the swamps, it still felt like an adventure to her. The land was welcoming but mysterious and also full of life.

Sara and Cyrus paddled down a back way lined with mossy mangrove trees. As a kid, Sara used to imagine the trees were giant, long-haired witches welcoming them to the swamp. The siblings looped around the watery cul-de-sac, passing a turtle on a log. When Cyrus and Sara were younger and went fishing with their dad, they used to call this part of the bayou the "fish neighborhood." They always got their best catches there.

Wetlands

A wetland is an environment where land is covered by water a significant amount of the time. Wetlands can be near the sea, near rivers and lakes, or in brackish water where fresh and salt water mix. Marshes, swamps, bogs, and fens are all types of wetlands. A bayou is a name for a slow-moving body of water. The word is most often used in the southeastern United States, especially in Louisiana where there are many bayous. Wetlands thrive around these bayous.

THE WARNING

Sara and Cyrus headed back into the main part of the bayou and lazily paddled south. Cyrus was in college studying environmental science. He told Sara about how swamps such as this one used to cover much more of the land in Louisiana. The rivers and the seas had risen and receded over thousands of years.

"These swamps have always flooded and dried out," Cyrus said. "But when the European settlers came and built the dams, we lost some swamps forever. And the more people build, the more swampland we lose."

"OK, professor," Sara said, pretending to roll her eyes. Secretly, she was fascinated. She tried to imagine what this swamp would have looked like thousands of years ago.

The siblings kept an eye out for the brown snout of an alligator in the water. The giant reptiles sometimes popped up in the water. But Sara saw something else instead. A sign stuck in the mud said, "Land for Sale," and then "Do Not Enter."

Sara was surprised. For as long as she could remember, Saulnier Swamps and the surrounding wetlands had been free for everyone to use. It was an unofficial agreement between the family and the community. People could camp, boat, and fish there as long as they treated the land and the wildlife with respect. But now, it looked like things might be different. Cyrus and Sara didn't want to get in trouble, so they cut the trip short

American alligators can grow more than 12 feet (3.7 m) long.

and paddled back. The sun was high, and the mosquitoes were already out, but Sara still wished they could have paddled for longer.

WHO OWNS THE SWAMP?

Once she was back at the house, Sara got on the computer and searched for *Saulnier Swamps*. After a couple of minutes of digging, she saw that the Saulniers were leaving the coast and selling the land around the bayou. A chain restaurant company wanted to build a seafood restaurant right in the middle of it.

Sara couldn't believe it. What would happen to the nature she'd grown up with—the turtles, the alligators, and even the mosquitoes? She looked online and tried to figure out what happens to swamp animals when their habitats get turned into places for humans. She knew that some animals could live alongside people, such as the raccoons she sometimes saw scurrying around the dumpsters in town. Others, she learned, depended on the bayou to survive.

Sara read about the Gulf sturgeon, a giant fish that

Wading Birds

Herons, egrets, ibises, cranes, flamingos, and storks are all examples of wading birds. As their name suggests, wading birds are types of birds that walk, or wade, in shallow water. This is how they find food such as fish or lizards. These birds have certain features in common, including long legs, beaks, and necks. Wading birds live in or near bodies of water, and many species can be found in both fresh and saltwater habitats.

More than 2,000 square miles (5,180 sq km) of land in Louisiana have been lost since the 1930s, with rising sea levels being a major cause of that loss.

can grow up to nine feet (2.7 m) long and travels from the sea to inland waters to lay its eggs.[1] Gulf sturgeons were endangered because they had lost much of the habitat they depend on. Sara also read that the whooping crane, a big, beautiful bird that lived in and hunted in the marshland, disappeared completely from Louisiana after losing its habitat. Sara didn't want that to also happen to the long-necked herons and ibises she usually saw in the swamps.

Cyrus brought Sara coffee and listened to her explain about the seafood restaurant, including the muddy land that would need to be filled to make a parking lot. "Once that land is filled, there's no going back," Sara said. "And where do the animals go after that?"

Cyrus nodded, his face serious. "It's a big problem," he said. "And it's not just that. Remember when I said the waters in this area have always flooded, then dried out? Well, lately the water has only been rising. Not just here in Louisiana, but all around the world.

"The ocean is warmer than it used to be," Cyrus continued. "And warm water expands, meaning it increases in volume. Not just that, but ice caps are melting. So there's more water in the ocean. And that water is covering more and more land along the coast as sea levels rise."

Human activities and settlements can disrupt ecosystems by causing habitat loss or polluting the environment.

"Is that such a bad thing?" Sara asked. "Maybe the sea levels rising would make people think twice before building new seafood restaurants."

Cyrus smiled at Sara. "Yeah, but that means the swamps don't just get more water. They'll be completely underwater. And more salt water mixes with the fresh water on land. That kind of water mixing can be dangerous. It can throw off a very delicate balance, one that the wildlife depends on."

Rising Sea Levels in the United States

Across the United States, the sea level rose an average of 6.5 inches (16.5 cm) between 1950 and 2016.[2] This sea level rise is especially high on the Gulf Coast, which Louisiana shares with Texas, Mississippi, Alabama, and Florida. Louisiana has experienced about two feet (0.6 m) of sea level rise along its coast.[3] This means land that used to exist on the coast is now underwater. In addition, rising sea levels mean more frequent flooding for coastal communities, as well as more damage to wild habitats. Scientists estimate that between 2020 and 2050, the sea level along the US coast will rise an average of ten to 12 inches (25–30.5 cm).[4]

TAKING A STAND

As Sara lay in bed that night, she thought about the swamp and what Cyrus had said. She was scared for the land, for the animals, and for herself. But she didn't want to feel hopeless. Sara thought about her civics report that she had been putting off for days. The report was about how someone can call their local government officials and talk to them about important issues. Sara knew exactly what she was passionate about.

The next day, Sara read online that it was possible for local governments to buy land that was for sale. If the land is for the public to use, then local governments can purchase and manage it. Governments can also work with landowners to conserve land so that the habitats and wildlife inside of it are protected.

Sara knew just who to talk to about all of this. She called her friend Jermane, a guy who loved birds more than anyone she knew. He'd be furious if he knew that an important habitat for local birds might become a shrimp shack.

Two weeks later, Sara met Jermane outside city hall. They still had their posters from the rally they went to that morning near the entrance to Saulnier Swamps. "Wild Land Stays Wild" was printed on Sara's sign, and Jermane's read "The Birds Need a Home." They invited everyone they knew and spent the day raising awareness about the threat to the swamp, hoping they could get people to realize the special beauty of the wildland. Now, it was time for the next phase of action.

Sara read that there was going to be a hearing about the land sale at city hall. The restaurant owners needed special permission for their building plans. Sara and Jermane were allowed to speak at the meeting. They were going to make the case that the city, not the restaurant, should buy the land and preserve it as a home for important ecosystems.

Sara and Jermane had a speech prepared. Cyrus helped them write it before going back to school. Sara had never memorized so many statistics about frogs in her life.

Many activists protest to raise awareness about threats to the environment.

Sara took a deep breath and thought of the peace she felt when paddling through the swamp she loved. She was determined to muster all her bravery to help keep it safe. Sara imagined herself being cheered on by the mangrove trees as she walked inside the building.

Scientists conduct surveys to collect data on the conditions of different ecosystems.

CHAPTER 2

THE SCIENCE OF LAND PRESERVATION

Land preservation is the permanent protection of land from certain types of use. Preserved land is set aside from human development, such as building or mining. This land is often free from human use or management. If humans are allowed to use the land, they can use it only for specific purposes, such as for hunting or camping.

Land is often preserved for its natural aspects, but it can also be preserved for cultural or historical reasons. For example, Colonial Williamsburg is a preserved site in Williamsburg, Virginia. The site maintains buildings from early US history, educating the public about how some people lived during that time.

LAND AND LAND USE

Currently, only about 30 percent of Earth's surface is covered by land, a total area of about 57 million square miles (148 million sq km).[1] This land provides humans and millions of species of land-dwelling organisms with the space they need to live. About 19 percent of the land on Earth is considered wild, or

In 2022, about 13 percent of the land in the United States was permanently protected.[3]

unused by humans.[2] The rest is affected by human life in some way.

Humans may live, build, or farm on land, or they may extract resources such as oil or minerals from it. Land use happens on different scales, from light to intense. For example, some people in Europe, Asia, Africa, and South America practice a lifestyle called nomadic pastoralism. Pastoralists tend to live on rangelands, which are places where grasses are the main plants. They keep herds of livestock animals, such as camels or llamas, that they continually move from place to place. The animals graze on wild plants, and the people make temporary homes that they can destroy or move with them.

This lifestyle tends to work well with the natural ecosystems of rangelands. Pastoralist herds eat natural grasses and help spread plant seeds. In this and other ways, pastoralists can help preserve the biodiversity of a land area, maintaining its variety of life.

Other types of human land use, however, are more intensive. When humans make cities and towns, they significantly alter the way natural forces interact with the land. For example, when people pave over land with concrete to make buildings or roads, they create impervious surfaces. These artificial surfaces block water from flowing into the soil the way it normally would.

Land use also affects the types of wildlife that can thrive in an area. If a forest is cut down to make way for human settlement, the plants, animals, and fungi that depend on trees can no longer make a home there. Human water use can affect the distribution of land as well.

Some cities and towns are built on drained swampland. When the water from a wetland is diverted through draining or dams, the land changes. Wildlife that depended on the original ecosystem can no longer thrive there. This effect on land is called habitat loss. Human land use is the top driver of biodiversity loss worldwide.

Agriculture is another major type of land use. When farmers use land for crops, they decrease the natural variety of plants in an area. This leads to habitat loss for native plants and the wildlife that depend on them. Agricultural land use affects the way water and land interact as well.

For example, rice is a commonly grown crop. It is also semiaquatic, meaning that it must be grown in watery areas. Farmers must irrigate their rice fields to

Conservation versus Preservation

Land conservation and land preservation are related concepts. However, they have slightly different goals. The National Park Service explains that conservation "seeks the proper use of nature" while preservation "seeks the protection of nature from use."[4] In other words, land preservation has more of a focus on protecting areas, especially vulnerable ecosystems, from human use and development. Conservation, on the other hand, looks to find responsible ways to use natural resources to prevent damaging the environment. For example, a conservation project might try to find ways to sustainably cut down trees in an area, while a preservation project might try to keep the trees from being cut down altogether.

About 38 percent of land on Earth is used for agriculture.

provide the crops with enough water to survive. When water is used for growing crops, it can become polluted from substances such as fertilizer chemicals. A lack of clean water means that native plants are no longer able to grow, and plants, animals, and other organisms experience a disruption to their food and water supply. In addition, agricultural land use affects the soil that sustains land.

SOIL

Soil is the top layer of Earth's surface. It is a material that lies on top of land and has been exposed to weather and other natural processes. Soil is made of minerals and organic material, such as decomposing plants, mixed with water and gases. The major types of soil are dry, sandy soils; clay soils; silt soils, which store water easily; and loam soils, which are a mixture of sand, clay, and silt.

Soil differs from rocks, which are collections of minerals. Both soil and rocks are found on the surface of land, but soil plays an essential role in sustaining ecosystems. Most plants grow in soil. Plants take root in the soil and absorb nutrients from it. Fungi also live in soil. They help decompose dead things, turning the remains of life-forms into new types of matter. And animals such as worms make their homes in soil as well.

> **"Everything in nature—plants, animals, soil, water, air—operates in a web of interdependency; a tightly knit network of competition and cooperation.[5]"**
>
> *—Fred Kirschenmann and David Gould, conservation and agriculture activists*

Soil also helps distribute water around the world. Rainfall, for instance, is absorbed by soil and travels through it, often into sources of water such as rivers. This process is called infiltration. Biomes, or environments with specific climates and organisms, have different types of soils.

Human land use can degrade soil by stripping its nutrients, air, water, and organisms. For example, people need soil to grow crops such as wheat and corn. When farmers plant crops, they replace the diversity of natural plant life with just one or two types of plants. This can disrupt the balance of living things in the soil. The soil may lose key nutrients, and small organisms in it may no longer thrive. In addition, planting crops in an area such as a former forest can get rid of trees that once shaded the soil, potentially causing the soil to dry out and be worn away. This process is called erosion.

CLORPT

There are five major factors that affect how soil changes. Scientists use the acronym CLORPT to describe these factors. CLORPT stands for *climate*, *organisms*, *relief*, *parent material*, and *time*. Climate is the long-term weather pattern in the area where the soil is located. Organisms are living things, and relief is the formations and features of the landscape. Parent material is the different types of rocks that the soil comes from. Time is the way soil changes during its life span. For example, soil tends to weather over time. Older soil breaks into smaller pieces as a result of forces such as friction and wind.

LAND AND THE CARBON CYCLE

Animals stay alive through a process called respiration, also known as breathing. They take in oxygen, which is an element found in both air and water. When animals breathe out, they expel carbon dioxide. Plants survive through a similar process called photosynthesis. They take in carbon dioxide from the atmosphere and energy from the sun, and then they release oxygen into the atmosphere.

The carbon dioxide released by living things is stored inside plants and in the soil. It is also absorbed into the ocean. The process by which carbon moves through the atmosphere, ground, plants, and sea is called the carbon cycle.

Many environments both store and release carbon dioxide. A place that stores more carbon than it releases is known as a carbon sink. Forests tend to be carbon sinks because the trees and the soil store more carbon than other living things, such as animals, release. The ocean is another important carbon sink. About 25 percent of Earth's carbon dioxide is stored by the ocean.[6] Rocks and underground deposits also act as carbon sinks.

Researchers will sometimes take soil samples to assess the health of an ecosystem.

Although it's important for life, carbon dioxide is a greenhouse gas. This means that it helps trap heat inside the atmosphere. When the amount of carbon in the atmosphere increases, so does the temperature of the planet.

Since around the mid-1800s, the amount of carbon dioxide in the atmosphere has greatly increased. Average global temperatures have gone up as well. This is mostly due to human activity, including burning fossil fuels, such as oil and coal. Land use plays a role in the increase of atmospheric carbon. When forests disappear or soil is eroded, the planet loses a source of carbon storage. That means

During periods of drought, water sources can dry out, and soil can start cracking from a lack of moisture.

that the carbon once stored in plants or in the ground ends up in the atmosphere.

The increase in atmospheric carbon has changed the average temperature of Earth over the past 200 years. In the 2020s, scientists found the surface of the planet to be about two degrees Fahrenheit (1.1°C) warmer than it was 150 years earlier.[7] This increase in temperature is known as global warming, and it is predicted to continue over time. Global warming is one of the many by-products of the ongoing threat of climate change.

Climate change is the shift in global temperatures and weather patterns over a long period of time. It includes much more than just global warming. For example, higher global temperatures due to climate change cause moisture to evaporate into the air more quickly. In wetter regions, this evaporation makes severe storms such as hurricanes more likely. And in drier regions, it means that the already scarce water sources are more likely to dry up, leading to droughts.

In the past 200 years, the amount of carbon dioxide in the atmosphere has increased by more than 50 percent.[8]

Climate change affects land throughout Earth. It threatens the availability of water, the livability of the land for wildlife, and the quality of soil. In order to preserve land, it is necessary to understand how it has been affected by climate change and the challenges different environments are likely to face in the future.

HOW LAND PRESERVATION WORKS

There are a few types of land preservation. It is up to the people who own land to decide how it can and should be used. Some land is private, meaning it belongs to individuals or companies. Other land is public, belonging to a government body. In both cases, people use laws, contracts, and agreements to decide when and how to preserve land.

Scientists play an important role in deciding when and how land should be preserved. Conservation scientists are people who study the way land works. They work across a wide range of scientific specialties, such as biology, botany, and geology. For example, a conservation scientist might make a plan to study how climate change or other human activities affect preserved land and the species that live there. These scientists then help governments and organizations make decisions about how to properly manage the land and stop it from being further affected.

One common way to preserve private land is through a conservation easement. This is a legal agreement that specifies how a piece of private land can be used. In the United States, private landowners make these agreements with a government or trust. The agreements protect a piece of land from human development even after the original owner's death. These easements have some flexibility. They can allow for small amounts of construction or for certain activities such as cutting down some trees.

Rare Plants and Land Preservation

Land management and wildlife preservation are closely linked. For example, because different plants are native to certain areas of land, plant science plays an important role in land preservation. Some plants only grow in very small geographic areas, so in order to preserve the plant, the land it grows on must also be preserved. In the United States, the Bureau of Land Management helps document and maintain more than 1,800 species of rare plants.[9] They reserve some land as Research Natural Areas, places where human use is limited to scientists studying plants and other features. Other lands are called Areas of Critical Environmental Concern, where management focuses on maintaining and improving land for native species.

In the United States, federal, state, and local governments can all own and manage public land. Another form of US land is tribal land, which belongs to American Indian tribal governments. Types of public land in the United States include national parks and forests, monuments, and wildlife refuges. These public lands are managed by government agencies that oversee their proper use.

Countries across the world have different processes for preserving land. About 16 percent of land globally is considered conserved or protected.[10] Some organizations work to help countries come together and preserve areas that cross international borders. For example, the International Land Conservation Network is a nongovernmental organization (NGO) based in Massachusetts. The organization collects information on land preservation in different countries. It works to connect organizations working on land preservation worldwide so that larger areas of land can be managed and preserved across political boundaries.

Although most areas of Antarctica are highly protected, the continent does have some scientific research stations.

The remains of many former civilizations can be found around the world today, including those of the Inca Empire in Machu Picchu, Peru.

CHAPTER 3

THE HISTORY OF LAND PRESERVATION

Humans need land for shelter, food, and water. Since the beginning of the species, humans have made their mark on land. Archaeologists have found evidence of fires that early humans made approximately 1.5 million years ago. Scientists have also discovered signs of early human shelters that are about 400,000 years old.

Early humans rapidly spread across Earth. As they moved into new areas, their use of land affected the environments where they settled. For example, early humans in northern climates hunted large mammals, such as the woolly mammoth. When humans moved into areas where these large mammals lived, species extinction often followed. This pattern began as early as about 125,000 years ago.

Land and living beings are interconnected. Losses in biodiversity affect the land and vice versa. Large mammals, such as bison or the extinct woolly mammoth, help spread plant seeds and fertilize soil through their droppings. When a species disappears, the land is altered by the loss.

Land Use and Species Extinction

Sometimes, land use can lead to the extinction of certain species. This is likely to happen when a species depends on a particular land environment to reproduce or eat. Human land use has a long history of causing species extinctions. For example, the dodo bird was a large, flightless bird that used to live on the island of Mauritius and a few other islands in the Indian Ocean. In the 1600s, Dutch soldiers arrived to the island. The Dutch cut down the forests where the dodo lived and brought in animals, such as rats, that invaded the dodo's habitat and ate their eggs. These factors contributed to the dodo bird's extinction.

On the other hand, some types of early human land use helped ecosystems thrive. People have lived on the continent of Australia for more than 50,000 years. The Indigenous peoples of Australia's land are known as the Aboriginal peoples. Indigenous people are people who are the original or native inhabitants of a particular place and have a distinct culture and lifestyle.

Much of Australia's land is dry and grassy and is prone to wildfires. For thousands of years, people living on this land set small, controlled grassland fires. These fires served many purposes. One of their purposes was to clear away dead plants in order to make room for new ones to grow. In turn, this new growth would create habitats and feeding grounds for animals. This type of land use is an example of how humans can live on land in ways that help preserve it.

Some scientists have found evidence that people across the world have been helping preserve the land they have lived on for thousands of years. A paper published in 2021 estimated that by around 12,000 years ago, about three-quarters of the world's land was being used by humans.[1] The paper discussed that places with very old human settlements also had a high level of biodiversity. This suggests that in many places in the world, the presence of people and successful land preservation went hand in hand.

AGRICULTURE, CITIES, AND LAND PRESERVATION

Humans invented agriculture about 12,000 years ago. They discovered that they could domesticate wild plant species, turning them into crops. For example, scientists have found evidence that people in what is now the Middle East began planting fig trees 11,300 years ago. Planting crops in a concentrated area was easier and yielded more food than moving around to find wild plants.

Humans also found that they could change plants over time to make them into better food sources. Many common crops, such as corn, are the descendants of wild plants that were bred to be larger, better tasting, or easier to harvest. Along with plants, people began to domesticate animals, including goats and pigs.

In approximately 1000 CE, less than 4 percent of the world's usable land area was farmland. In 2019, that percentage had grown to about 50 percent.[2]

The invention of agriculture helped humans create larger, more permanent settlements. People had an easier time gathering large amounts of food, allowing them to stay in one place rather than constantly traveling. The first cities were created around 7500 BCE. These larger human settlements transformed wild tracts of land into places for human development and manufacturing. This activity altered natural ecosystems.

For example, cities built near rivers depended on freshwater sources for drinking, bathing, and irrigating crops. Human use of the water supply often meant less water for local plants and animals. The redirection of water flow and the planting of crops affected the soil in those areas as well. Researchers have found that the ancient city of Acre, in what is now the Middle East, radically changed the local land 6,000 years ago. A forest near

The seasonal flooding of the Nile River in Egypt made fertile land for growing crops.

the ocean became a grassland after people began overusing the water supply in the area.

Human warfare also led to the destruction or degradation of natural land. For example, salting the earth is the act of making land unusable for farming. The term comes from an ancient Mediterranean wartime practice. Soldiers would conquer a city and spread salt on the fields of the conquered land. They knew that the salt would throw off the delicate balance of the soil. The intent was to leave the land barren so that crops and other plants couldn't grow.

THE INDUSTRIAL REVOLUTION AND LAND USE

The Industrial Revolution was a major change in human societies. It began in the late 1700s with the invention of new machinery in Europe. Before the Industrial Revolution, most people lived in rural areas and made goods such as clothing by hand. The invention of machines, including the power loom, made it easier and faster for people to create common goods. People in Europe began migrating from rural areas into towns and cities to start producing goods in factories. These factories often ran on fossil fuels, releasing greater amounts of carbon dioxide and other greenhouse gases into the atmosphere.

The Industrial Revolution led to a huge increase in the world population. Between 1625 and 1825, the world population doubled from 500 million to one billion.[3] This surge in population led to an increase in land use as well. In 1700, half the land on Earth was considered wild and another 45 percent was minimally used. By 2000, only 25 percent of the planet's land was wild, while 20 percent was minimally used.[4]

POPULATION GROWTH AND LAND PRESERVATION

In the 1800s, people started settling in large areas in the middle and western parts of the United States. In the mid-1800s, some Americans believed in a concept called Manifest Destiny. This was the idea that land in the Midwest and West was destined to be developed by European Americans. Driven by immigration and the desire for new living situations and business opportunities, millions of people moved west and displaced Indigenous peoples.

> "I recognize the right and duty of this generation to develop and use the natural resources of our land; but I do not recognize the right to waste them, or to rob, by wasteful use, the generations that come after us.[5]"
>
> *—Theodore Roosevelt, former US president, 1910*

As these settlers moved west, they drastically changed the landscape. In many cases, people found ways to profit from the natural resources that came from the land. The gold rushes of the United States, first in Georgia and then California, happened during the 1800s. People moved into areas where gold had been found, hoping to mine the land and grow rich.

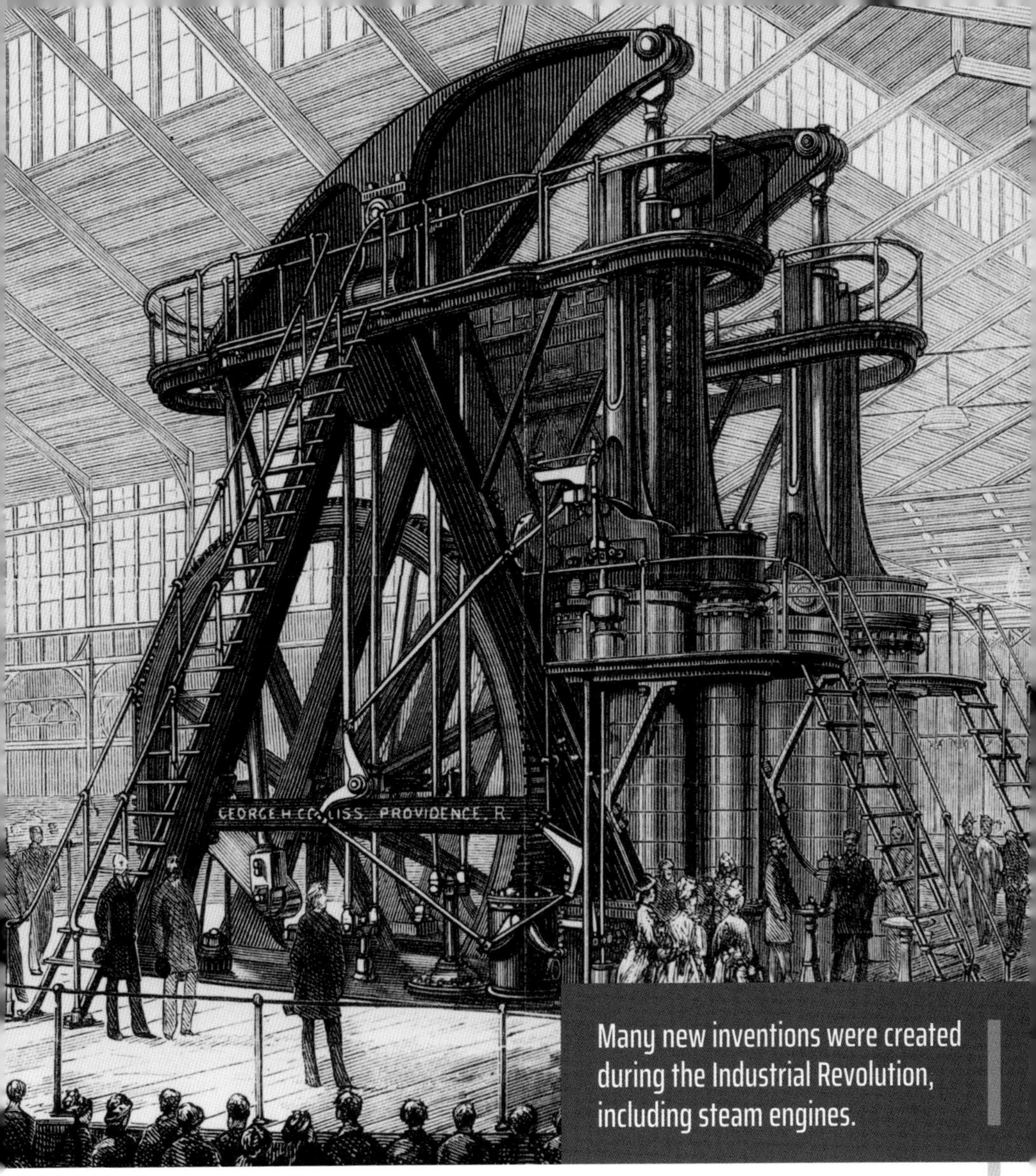

Many new inventions were created during the Industrial Revolution, including steam engines.

The surge in population in these areas happened quickly. For example, gold was discovered in California in 1848. By 1860, the state had a population of more than 300,000 people, almost three times what it had been in 1847.[6]

The Industrial Revolution also popularized new forms of transportation, including railways. In the 1800s, trains started connecting the eastern and western parts of the United States.

> **We owe it to ourselves and to the next generation to conserve the environment so that we can bequeath our children a sustainable world that benefits all.[7]**
>
> *—Wangari Maathai, founder of the Green Belt Movement*

These trains ran on steam created from the heat of burning coal. The construction of the railway had negative effects on land and ecosystems. For example, the American bison was once a key part of the US grassland ecosystem. The railroads opened the bison's habitat to settlers, and the animals were almost hunted to extinction.

During the mid- to late 1800s, as technological advancements were being made, people started becoming more aware that natural land was under threat. Many people criticized the attitude of Manifest Destiny, seeing it as a culture of wastefulness and disrespect for nature. It was during this time that some of the ideas of the modern conservation movement were born.

American nature writer John Muir advocated for wild US land to be preserved in public parks. Other public figures, such as architect Frederick Law Olmsted, who helped design New York City's Central Park, joined in the call to preserve parks. In 1872, Congress established Yellowstone National Park in Montana and Wyoming. Yellowstone became the first US national park.

President Theodore Roosevelt supported the conservation movement and helped expand the National Park System. He served as president from 1901 to 1909. In 1906, the US government passed the Antiquities Act. This act allowed the country to preserve places that are considered historically or archaeologically important. In 1916, President Woodrow Wilson signed the Organic Act to create the National Park Service. The act established that the National Park Service was responsible for maintaining current and future national parks and monuments, as well as other preserved land.

The US Bureau of Land Management was created in 1946 to help take care of public land. It currently manages 30 national monuments.[8] Some of these monuments are natural, such as the Grand Canyon. Others, including the Statue of Liberty, are human-made. Today, there are 428 sites in the United States managed by the National Park Service, with 63 designated as national parks. These sites can be found in all 50 states and some US territories.[9]

Wrangell–Saint Elias National Park and Preserve in Alaska is the largest US national park. It covers 20,600 square miles (53,400 sq km).

THE ENVIRONMENTAL MOVEMENT

In the early to mid-1900s, environmental activists began talking more about the importance of keeping land wild and conserving resources. A lot of art and literature celebrating wild landscapes was created during this time. It was also a time when many policies and acts related to the environment were put in place.

In 1935, the US Congress established the Soil Conservation Service, now known as the Natural Resources Conservation Service. This government agency was created to minimize soil erosion and help farmers manage land. Over time, it expanded to include many forms of land conservation assistance.

In the 1960s, the environmental movement caught on in the United States and abroad. At the time, many activists focused on issues such as pollution and saving wildlife. This focus eventually led to the creation of many international NGOs and the Endangered Species Preservation Act of 1966.

Marjory Stoneman Douglas and the Everglades

Marjory Stoneman Douglas was a writer and activist. She lived in Florida and became interested in the Everglades. The Everglades is a large series of wetlands in southern Florida. It stretches more than 3,100 square miles (8,000 sq km).[10] Starting in the 1800s, humans began to develop the Everglades for agricultural and urban use. Douglas realized that many people did not appreciate the importance of the Everglades to wildlife and the water supply. In 1947, she wrote a book called *The Everglades: River of Grass*. The book changed public opinion about the importance of preserving wetlands. That same year, the Everglades officially became a national park.

Land preservation found a new approach in 1985 when American activists Randy Hayes and Mike Roselle founded the Rainforest Action Network. Hayes and Roselle thought of new ways to get people interested in preserving forested land. They popularized using the term *rainforest* instead of jungle.

Captain Planet and the Planeteers

Captain Planet and the Planeteers was an animated TV show that ran from 1990 to 1996. The show was about a group of five teenagers who had special rings they could combine to summon Captain Planet, a superhero who helped save the environment. Together, the teens and Captain Planet fought villains who wanted to harm the planet, including those damaging land. In 1991, the creators of the show established the Captain Planet Foundation to encourage young people to get involved in environmental activism.

The Rainforest Action Network and other movements made the connection between deforestation and the clearing of land for crops. Hayes and Roselle encouraged people to boycott companies that cleared rainforest land. They also helped popularize the "Save the Rainforest" movement through outreach efforts, such as TV advertisements and T-shirts.

In other countries, activists began leading movements to fight deforestation and promote healthy land. For example, Kenyan activist Wangari Maathai founded the Green Belt Movement in 1977. Maathai recruited women to plant trees in degraded areas of land in Kenya to help prevent soil erosion and forest loss. She won the Nobel Peace Prize in 2004 for her efforts.

Boreal forests are also known as taigas. Russia, Canada, and the United States are home to most of the world's boreal forestland.

CHAPTER 4

FOREST PRESERVATION

A forest is a place where trees are the dominant plant life-form. Trees are woody plants that are perennial, meaning they grow from year to year. In general, trees have one stem, grow leaves, and grow to be at least 13 feet (4 m) tall.[1]

There are a few major types of forests around the world. Tropical forests are located near the equator. They have a heavy forest canopy, with little light reaching the ground. Temperate forests are found farther from the equator in climates with multiple seasons. They tend to be dominated by deciduous trees, or trees that shed their leaves.

Both temperate and tropical forests can be rainforests, or forests in areas with heavy rainfall. Boreal forests cover about 11 percent of land and are located in the northern part of the world at high altitudes. These forests are dominated by conifers, which are trees such as pines that have cones and needles.[2] Together, about 30 percent of the land in the world is forested.[3]

Forests are home to the biggest variety of life-forms on Earth. About 60,000 different kinds of trees can be found in forests.

Forests also host the majority of the world's mammal, bird, and amphibian species. About 1.6 billion people around the world rely on forests for their livelihoods, including for food, shelter, and jobs.[4]

Forests play a vital role in the carbon cycle. They act as carbon sinks, absorbing carbon from the atmosphere and storing it in the trees and soil. Scientists estimate that around 45 percent of the world's carbon is stored in forests.[5] But when forests are cut down, some of that carbon returns to the atmosphere.

CAUSES AND EFFECTS OF DEFORESTATION

Deforestation is the act of clearing forests to make way for human activities. People sometimes clear forests to create space for buildings and settlements. But oftentimes, the major driver of deforestation is agriculture.

The majority of agricultural land in the world is used for livestock. These animals, especially cows, require a lot of land. Some crops, such as soy, need to be grown just to feed them. In addition, livestock need room to live and move. Even though there is a lot of work that goes into raising livestock, it is still a profitable industry, especially beef. The clearing of land to grow crops and raise livestock can be devastating for wildlife and can contribute to greenhouse gas emissions.

Forests are home to 80 percent of the world's plant and animal species that live on land.[6]

The Amazon rainforest is the largest rainforest in the world. The forest is located in South America and stretches across 2.2 million square miles (5.7 million sq km) of land in eight countries and one territory, with most of its area in Brazil.[7] The Amazon contains a vast amount of biodiversity. It is estimated that 10 percent of the world's species live inside the Amazon rainforest.[8] The forest may be home to up to 400 billion trees, and scientists estimate that it stores about 150 to 200 billion short tons (136 to 181 billion metric tons) of carbon.[9]

Palm Oil and Deforestation

Palm oil is made from the fruit of oil palm trees. These trees grow as crops in warm-climate countries such as Indonesia. Palm oil has no smell or taste, and it helps keep food from going bad. For this and other reasons, it is an extremely popular product ingredient. In fact, palm oil is in almost half of all packaged products at the grocery store. However, palm oil crops are grown in the areas where tropical forests grow. The European Commission estimates that this crop alone contributes to about a quarter of deforestation.[12]

However, the Amazon is rapidly losing territory. Since the 1970s, close to 20 percent of the rainforest has been destroyed.[10] Most of this destruction is due to deforestation for agricultural practices, especially cattle ranching. When the rainforest loses territory, wildlife lose their habitat and carbon is released.

But deforestation is not unique to the Amazon. Between 10 and 20 percent of the global increase in carbon in the atmosphere is due to deforestation.[11] Many forests are threatened by agriculture and industry. For example, the rainforests of

Deforestation can cause habitat fragmentation, which is when large land areas are spilt into smaller, disconnected areas, often due to human activities.

central Africa are home to deposits of minerals such as gold. Mining operations use fossil fuels and release greenhouse gas emissions, and the runoff from mining and refining operations often pollutes water. These mining operations can displace and harm wildlife.

In 2022, several NGOs spoke out against what they believed was an illegal deal between the Democratic Republic of the Congo and a Chinese mining company. The NGOs claimed the mining company opened a gold mine in what should have been preserved lands. These conflicts between companies, activists, and governments can be difficult for advocates of land preservation to navigate.

ACCELERATION OF DEFORESTATION

Deforestation is a process that, if not interrupted, tends to get faster over time. As a biome becomes less diverse, natural processes, such as the decomposition of organic material into soil, are weakened. As a result, soil and water quality tend to decline. This means that even land that is not heavily used can suffer long-term effects from biodiversity loss.

Deforestation also increases the risk of other hazards that threaten forests and people. For example, the deforestation of the Amazon rainforest reached a peak in the early 2000s. To help decrease the level of deforestation in the Amazon, the government of Brazil, which is where most of the forest is located, has planned to end deforestation by 2030. But even with efforts to combat the loss of

Deforestation and Disease

Zoonosis is any infection or disease that spreads from animals to humans. This can happen when humans touch or eat an infected animal. It can also happen when humans and animals share resources, such as a water supply. Zoonotic diseases include HIV infection, Ebola virus disease, and salmonella infection. About 60 percent of known infectious diseases are zoonotic.[13] Deforestation can make the spread of zoonotic diseases more likely. When wildlife lose their forest habitats, they are more likely to live near humans and share water and land with them. This closer contact means a higher risk of new diseases passing from wildlife to people.

trees, the risk of wildfires in the rainforest has increased. As more of the previously forested land becomes drier shrubland and grassland, forest fires become more likely.

One fire hazard in the Amazon is slash-and-burn agriculture. Slash-and-burn agriculture is when people cut down trees in a forest, burn them, and use the ash to fertilize soil. The soil that's left from slash-and-burn practices is often usable for only a few years before it becomes degraded. This is because the layer of

To help manage and put out wildfires, specialty planes will release large amounts of water over burning areas of land.

ash that covers the soil runs out of nutrients in a shorter amount of time. Without this layer, and without an ecosystem of diverse plants to nourish and anchor it, the resulting soil erodes easily and cannot sustain crops. In addition, the fires used in slash-and-burn agriculture can get out of control, burning more land than people expected.

> "[Rainforests] are in fact the world's air-conditioning system—the very lungs of the planet . . . and we are on the verge of switching it off.[16]"
>
> *—Charles, Prince of Wales, 2008*

Climate change has also caused the Amazon rainforest to become hotter and drier. This increases the risk of bigger, longer-lasting wildfires. In 2019, there were 75,000 fires in the Amazon, which was an 85 percent increase from the previous year.[14]

Some scientists believe that the effects of deforestation and climate change will continue affecting forests past a point of no return. They think that the forest may not recover from deforestation and climate change, even if humans stop cutting down trees and work to help restore ecosystems. In 2020, geographer Robert Walker predicted that climate change would cause the Amazon rainforest to transform from forest to grassland by 2064. "If southern Amazonia's dry season continues lengthening," Walker wrote, "the drought of 2005 will become the region's new normal."[15] Not all scientists agree with Walker's prediction. However, many do agree that ecosystems such as the Amazon can reach a tipping point from which they cannot recover.

INDIGENOUS PEOPLE AND FOREST PRESERVATION

Indigenous people have a large and difficult role to play in the preservation of forests. About 80 percent of the world's biodiversity is contained in Indigenous lands.[17] These areas include biomes such as forests, drylands, and wetlands. Many Indigenous peoples rely on preserved lands to maintain traditional ways of life.

Indigenous people that live in or near forests often rely on local plant and animal life for food, medicine, and other essentials. Their traditional ways of life work alongside effective forms of forest management and help protect forest land from degradation. For example, the Cofán people in the Amazon rainforest in Ecuador watch the forest animals and plants closely. They make community decisions about which species to hunt and harvest and which to leave alone.

On the Asian island of Borneo, Indigenous farmers manage tree farms with extremely high levels of biodiversity. They can manage the human-made forests and the biodiversity by paying close attention to the needs and behavior of the many different types of trees. Research shows that the rate of deforestation in forests managed by Indigenous peoples is 26 percent lower than in forests that are protected by governments.[18] This vast

Many Indigenous communities around the world fight to protect sacred lands from being used for fossil fuel operations.

responsibility of protecting land has led the United Nations (UN) to call Indigenous peoples "the unsung heroes of conservation."[19]

Nemonte Nenquimo is an Indigenous activist working to fight climate change and preserve land for traditional use. Nenquimo is a member of the Waorani people of Ecuador. Nenquimo's people are indigenous to the Amazon rainforest.

In 2015, Nenquimo cofounded an organization called Ceibo Alliance to fight plans to extract oil in her native lands. She was also elected president of an organization that represents the Waorani peoples of the Pastaza province in Ecuador. In 2020,

Nenquimo won a lawsuit that would help protect 780 square miles (2,020 sq km) of land from oil companies.[20]

SUSTAINABLE AGRICULTURE, REFORESTATION, AND PRESERVING FORESTS

Many land preservation activists seek better, more responsible ways to manage farmland. The sustainable agricultural movement's goal is to minimize the negative effects of agriculture, including soil degradation and erosion. One strategy is the practice of shifting cultivation, or leaving some land uncultivated while other land is used. This strategy gives soil time to restore its nutrients before being reused later for agricultural purposes.

Shifting cultivation can help even slash-and-burn agriculture be more environmentally sustainable. Some Indigenous farmers in the Amazon use slash-and-burn agriculture to cultivate small plots of land until they are overrun by native plants. Then, as the forest environment returns, they move to new farmland. Rotating and covering crops as well as mixing crop and livestock land can also protect soil. While these strategies can reduce the harmful effects of agriculture, they don't ultimately solve the problem of the clearing of new land.

Another potential solution to deforestation is its opposite, which is reforestation. Reforestation typically starts with people planting trees. The reforestation movement seeks to plant trees worldwide to combat the effects of deforestation. Tree planting can help replace lost forest cover and add more carbon storage capacity.

There are many organizations worldwide that focus on planting trees. For example, One Tree Planted is a nonprofit organization that allows people to donate to reforestation efforts across the globe. By 2024, the organization has planted more than 135 million trees.[21]

Scientists and students at the Forest Restoration Research Unit at Chiang Mai University in Thailand work to restore tropical forest ecosystems in the country.

Other strategies to protect forests include forest restoration. This approach focuses on all aspects of a lost forest biome. Forest restorers work to plant a diverse array of trees in an area. They clear underbrush, or shorter plants, in order to give the trees the best opportunity to grow.

Another strategy for recovering forests is natural generation. This strategy tries to create the best possible conditions for new trees to grow. For example, oak seeds will sprout if a forest fire burns off enough space for a seedling to take root. Natural generation measures include setting controlled fires to make room for those seedlings to grow. In the past, natural regeneration has been the main forest recovery strategy in the United States. However, some forests are so degraded that they need more active tree planting in order to recover lost territory.

Artificial Forests

Artificial forests are places such as tree farms that fit the definition of a natural forest but are planted and maintained by people. Some artificial forests can help restore soil and support wildlife. However, these human-made forests lack the biodiversity of natural forests. Land is often cleared to make artificial forests. In addition, crop trees are usually cut down and replanted. This means that, overall, some artificial forests may release more carbon than they take in.

LESEIN MUTUNKEI

Lesein Mutunkei is an activist from Kenya who loves soccer and the planet. Kenya is suffering from widespread deforestation. Tree cover in the country decreased by 11 percent between 2001 and 2022.[22] Mutunkei decided he would use his love of soccer to combat deforestation and climate change.

In 2018, Mutunkei started a program called Trees4Goals to get young athletes to plant a tree every time they scored a goal. But he decided to increase his own target to planting 11 trees for every goal he scored. Mutunkei focuses on planting trees that are native to Kenya. By 2022, he and his teammates had planted more than 5,500 trees. Mutunkei's commitment has inspired other Kenyan sports teams to do the same. By 2023, 15 school and club soccer teams from Kenya were participating in the Trees4Goals program.[23]

Mutunkei loves using the popularity of soccer to help spread his message. He said, "[Soccer] has the power to help us create global environment awareness and take global action to achieve a greener future."[24] Eventually, Mutunkei hopes to get the Fédération Internationale de Football Association (FIFA), which governs soccer around the world, involved in his tree-planting mission.

In 2020, Lesein Mutunkei won an International Young Eco-Hero Award for his environmental activism.

Wetlands cover about 6 percent of Earth's surface and are found on every continent except Antarctica.

CHAPTER 5

WETLAND PRESERVATION

Wetlands play important roles in the carbon cycle and the water cycle. These ecosystems have high biodiversity and provide homes for migrating wildlife. They also play important roles in the life cycles of animals such as amphibians, which must lay eggs in bodies of water.

In addition, wetlands have a high level of productivity. This means that plants grow quickly there. But human development threatens to alter or destroy wetlands, in some cases permanently. After many years of damaging wetlands, governments and organizations are looking for ways to protect them.

DEVELOPMENT, TRANSPORTATION, AND WETLAND PRESERVATION

When Europeans first settled in what is now the United States during the 1600s, they considered wetlands inconvenient and unnecessary. People disliked having to travel around marshes and bogs instead of through them. Colonists and settlers drained many large native wetlands in order to make room for roads, farms, and passageways. They did this by diverting the water away from wetlands to other places. This trend of draining and destroying wetlands continued in the United States through the 1900s.

The invention of the automobile helped lead to the phenomenon of urban sprawl. Urban sprawl is the rapid geographic growth of a city or town. In the United States in particular, urban sprawl accelerated after World War II (1939–1945). People began to live in highly developed suburbs that spread across large areas of land. Suburban developments, such as shopping malls and new neighborhoods, were built on top of wetlands.

Natural wetlands can purify water by removing up to 60 percent of metals and 90 percent of nitrogen from water.[2]

Throughout the 1950s and 1960s, the US government made little effort to protect wetlands. Some US policies basically encouraged wetland destruction. In the 1930s, the US government helped farmers drain wetlands for free so the areas could be used for agriculture. Between the mid-1950s and mid-1970s, the United States lost about 860 square miles (2,230 sq km) of wetlands every year.[1]

By the 1970s, the modern environmental movement and activists such as Marjory Stoneman Douglas had raised awareness about the importance of wetland environments. The US government began to take wetland preservation more seriously. In 1986, the Emergency Wetlands Resources Act

Wetlands are sometimes filled to create solid ground so highways, houses, or businesses can be built.

was passed. This act requires that the government make conservation plans for wetlands and map them.

However, the loss of wetland habitats has not ended. Scientists estimate that US states, excluding Alaska and Hawaii, lose more than 125 square miles (324 sq km) of coastal wetlands each year.[3] Human development is considered the main cause of wetland loss in the United States. However, rising sea levels and erosion also play a big role. This trend of wetland destruction is happening worldwide as well. Between 1900 and 2017, about 64 percent of the world's wetlands were lost.[4]

THE WATER CYCLE AND CLIMATE CHANGE

The water cycle is the process by which water moves through Earth's surface and its atmosphere. More than 97 percent of the world's water is found in the ocean and inland seas, and about 2 percent is frozen in glaciers.[5] The remaining fresh water on Earth is what humans and other land wildlife depend on for drinking, irrigation, and other essential needs.

The Pantanal in South America is the world's largest tropical wetland. It is home to more than 4,700 species of plants and animals.

Wetlands help to clean and purify fresh water that flows through them, removing impurities such as sediment. The water in a wetland is slow moving, so it tends to stay long enough that the impurities can be filtered out. This filtering happens through complex processes involving the physical flow of the water and interactions with wetland wildlife. After being filtered by a wetland, purified water then enters another body of water,

such as a river. Wetlands are sometimes called "nature's kidneys" because of their role in purifying fresh water, similar to how kidneys filter blood in the human body. When wetlands are lost, this water purifying function is lost as well.

> **"The challenge most commonly faced regarding [wetlands] and other protected areas is how to manage them. Therefore, the importance of wetlands needs to be promoted in order to get people to invest in conservation.[6]"**
>
> *—Martha Rojas Urrego, secretary general of the Ramsar Convention on Wetlands, 2017*

Human pollution from sewage, pesticides, or waste from factories can enter a body of water and make its way into wetlands. In large amounts, these pollutants may become too much for the wetland ecosystem to handle. For example, in the North Island of New Zealand, pollution caused a chain reaction that severely damaged a freshwater wetland. Chemicals from fertilizer entered a lake, increasing the level of nitrogen in the water.

Algae thrived on the increased nitrogen and grew rapidly in the lake, creating an algae bloom. The algae drew oxygen from the water. The lower oxygen levels, along with heavy rainfall and high temperatures, caused a large number of fish to die. The dead fish increased protein in the environment, and a deadly toxin called botulinum flourished on the protein. As a result of the botulism outbreak, birds who touched or drank the water became poisoned. Thousands of birds began to die as well.

Scientists measure the water levels in wetlands to monitor the health of the ecosystem.

Climate change is another big threat to wetlands, especially coastal wetlands. As glaciers melt from rising global temperatures, sea levels rise. Rising sea levels can flood wetlands near the coast. They can also mix seawater with fresh water. An increase in the salinity, or salt content, of freshwater wetlands can harm or kill wildlife.

Climate change also increases the likelihood of severe weather events such as floods and hurricanes. These weather events can physically damage wetlands. They can also introduce too much seawater into a wetland environment. Climate change has made flooding more common, but droughts are also more frequent than they used to be. This is because the warming of the planet has changed where and when rainfall travels. Both extremes can disturb wetland biomes.

WETLANDS AND THE CARBON CYCLE

Wetlands have some unique advantages when it comes to collecting and storing carbon. The first is that wetland plants tend to grow quickly. This means that they can capture a lot of carbon in a short amount of time. The other is that the soil in wetlands tends to lack oxygen or have very little of it. Oxygen is what helps organic material such as dead plants and animals decompose and break down.

This decomposition process releases carbon into the atmosphere. However, when material does not break down completely, carbon stays in the soil instead of being released. This means that wetlands can store carbon for very long periods of time.

Peat

Peat is a special type of soil layer that is made when decomposing plants and other matter become soaked in a watery environment. Peatlands are located mostly in the far northern parts of the world, but there are also some in Southeast Asia, northern South America, and parts of central Africa. Peatlands lack oxygen, the element that helps organic matter decompose completely. Without oxygen, this partially decomposed matter becomes a dark brown, dirt-like substance. It takes about ten years for 0.4 inches (1 cm) of peat to form.[7] Peat that is compressed and heated eventually becomes coal.

For example, peat bogs are wetlands that have a lot of peat, or partially decomposed plant matter. Some of the largest wetlands in the world are peat bogs. These bogs are excellent carbon sinks. Carbon can stay inside them for thousands of years. Just 3 percent of the world's land is peat bogs. However, they store twice as much carbon as all the forests in the world.[8] For this reason alone, preserving peat bogs is an important goal in the fight against climate change.

Sponge Cities

Urban architects and planners are looking for solutions to flooding in human settlements. One such solution is the creation of sponge cities. In China, people are facing two sides of a water crisis. On one hand, fresh, drinkable water is scarce. On the other hand, floods and heavier rainfall are happening more often. A "sponge city" refers to an urban place that has modified its infrastructure to help it better absorb water to prevent flooding. Some Chinese cities have made themselves "spongier" by planting trees and grasses in previously paved areas and making wetland parks. These areas inside the city help manage flooding.

SUSTAINABLE DEVELOPMENT AND WETLAND PRESERVATION

About 75 percent of the wetlands in the continental United States are privately owned.[9] This can make it more difficult for the government to set standards for wetland management. But there are still ways to develop human settlements so that wetlands and their wildlife can remain protected.

An important step to protecting wetlands is to acknowledge the connections between them and other types of habitats. Sometimes, it is not enough to preserve just one area of land.

Many people in Indonesia have taken part in restoring the country's mangrove forests by planting mangrove saplings.

The construction of highways, power lines, and other development projects can separate interdependent ecosystems from one another. This can make it difficult for wetland wildlife, such as amphibians and migratory birds, to follow their natural rhythms. It can also complicate the natural purifying function of wetlands. This segmentation of wildlife environments is called habitat fragmentation.

One way to address this problem is by creating wildlife corridors. Wildlife corridors are strips of land that connect different wild areas, making contact and transit safer and easier for wildlife. These corridors can be created alongside human development projects.

A 2022 study published in the journal *Sustainability* found that when areas around power lines were not mowed and were allowed to grow freely, they tended to host more wildlife than when they were mowed. This was also true for other suburban areas, such as backyards and lawns. The power-line corridors were found to have greater numbers of insect pollinators such as bees, whose movements across land areas are vital to the plant species they pollinate. Other animals use these corridors for safe passage from one wild habitat to another.

Wildlife corridors are also important for migratory species, especially birds. In the Americas, scientists have identified four major "flyways" that migratory birds use to travel between seasonal habitats. For example, the sandhill crane travels to the northern United States and Canada each spring, finds places to breed, and then makes the long journey south to the wetlands of Texas, Mexico, and Florida in the fall. To help preserve the corridors of birds such as the sandhill crane, the American Bird Conservancy creates BirdScapes, which are partnerships with public and private landowners to make safe habitats for these animals.

The international community has also begun to further address threats to wetlands. In 2022, the UN Biodiversity Conference, an international gathering designed to promote biodiversity, made an agreement on future steps to protect the environment. The agreement acknowledged that wetland loss worldwide has been accelerating. It also stated that wetlands have faced more biodiversity loss than other kinds of ecosystems.

Those involved in the agreement are committed to restoring 30 percent of degraded lands, inland bodies of water, and coastal and marine ecosystems. Leticia Carvalho, who is the head of the Marine and Freshwater Branch at the UN Environment Programme, said, "We're running out of chances to protect the services provided by wetlands that societies depend on for a sustainable future. . . . We must ramp up international solidarity, capacity-building, and funding without further delay."[10]

The Mangrove Breakthrough

Mangrove trees, which grow in and along wetlands, help store carbon. They also help keep carbon and wetlands in place by anchoring the carbon-storing soil. The Mangrove Breakthrough is a collaboration between a nonprofit called the Global Mangrove Alliance and members of the UN's climate change leadership. The goal of the Mangrove Breakthrough is to stop the loss of mangrove trees and restore habitats that have lost mangroves. The project has recruited other organizations and private companies to get involved in helping mangrove forests.

In the United States, government bodies such as the Environmental Protection Agency (EPA) have recognized the important role that wetlands play in the water supply. The EPA has developed strategies for monitoring the health of wetlands. It also offers training in watershed management for people who help oversee wetlands. This can help private owners of wetlands learn how to effectively take care of the land.

In 2000, the US Congress approved a project to help restore natural water flows in Everglades National Park.

The Sahara is the world's largest desert, covering about 3.3 million square miles (8.5 million sq km) of land.

CHAPTER 6

DRYLAND PRESERVATION

A dryland is a biome where water is scarce. Deserts are one of the most well-known drylands. A desert is a biome that receives less than ten inches (25 cm) of rain each year. Major deserts around the world include the Sahara Desert of North Africa, the Arabian Desert of the Arabian Peninsula, the Gobi Desert of central Asia, the North American deserts in Mexico and the western United States, and the desert interior of the Australian continent. About 20 percent of Earth's land is desert.[1]

Other types of drylands include grasslands, such as prairies and savannas. A prairie is a large, flat ecosystem that has some rainfall and very few trees. Prairies have mainly grasses and are usually categorized as tallgrass, midgrass, or shortgrass.

Savannas are ecosystems that are usually found in tropical and subtropical places near Earth's equator. They are often characterized as something between a forest and a desert because of their hot climate, lower levels of rainfall, and scattered tree cover. These ecosystems also have a layer of grasses on the ground.

In general, the world's large deserts are surrounded by other dryland biomes. About 42 percent of the world's surface is covered by drylands.[2] Even though drylands tend to lack trees, they are biodiverse environments. And more than two billion people live on drylands, making them important habitats for humans as well.[3]

DRYLANDS AND THE CARBON CYCLE

Like forests and wetlands, drylands play an important role in storing carbon. Drylands tend to have scattered or few trees, especially compared with forests. Severely dry areas do not support tree life well, while wetter areas may have some trees. For example, the drylands of Africa have about ten billion trees in total.[4] The Amazon rainforest, in contrast, is home to about 400 billion trees.[5]

But like wetlands, drylands store carbon underground and inside the soil. They also store carbon in other plants, including shrubs and grasses. Scientists estimate that drylands store about 30 percent of the world's carbon.[6]

Some of the carbon stored in dryland soil can react with calcium and turn into crystals. These crystals can be stored deep underneath the soil. When plant cover is lost and the soil erodes, this carbon is then released. In addition, when soil erodes, the tiny organisms inside it, such as mosses and bacteria, disappear as well. These living things in desert soil are known as the biocrust. They help the desert soil store carbon.

DESERTIFICATION

Desertification happens when drylands become degraded, losing plant cover and soil richness. As drylands lose soil and plant life, they become more desertlike and less viable for growth. These pieces of land do not become healthy deserts with unique biomes. Instead, areas affected by desertification can no longer support the biodiversity of their native biomes. Because Africa and Asia have the biggest dryland biomes, they are especially vulnerable to desertification.

About 47,000 square miles (121,000 sq km) of land is affected by desertification, land degradation, and drought every year.[8]

Desertification is a major threat to drylands. As a result of climate change, droughts have become longer and more intense in many places around the world. Drylands can cope with a lack of water, but prolonged droughts may disrupt the ecosystem's balance.

Agriculture can also cause desertification. Crops degrade drylands by depriving them of nutrients and making erosion more likely. In addition, the water used to irrigate dryland crops leaves salts behind in the soil. Few plants can grow in this salty soil, and land irrigated in this way may quickly become too degraded for agricultural use.

Population density can cause desertification as well. For example, almost two-thirds of Africa's drylands are home to

In 2022, scientists estimated that more than 27 percent of China was affected by desertification.

two-thirds of the continent's population.[7] Dense populations can stress land, depleting it of natural resources such as plants more quickly than they can be restored.

Desertification also affects food and water access. People in these areas are more likely to suffer from famine. If they lack access to clean water, they are more vulnerable to diseases transmitted through water and food. Soil erosion from desertification can also cause dust storms, which can pollute the air that humans breathe. The fight to preserve drylands from desertification is also a fight for human health and safety.

EXTRACTION AND DESERT LAND PRESERVATION

Oil and mineral extraction is a big threat to major deserts such as the Sahara. Oil is a fossil fuel. It was formed over the course of millions of years after tiny organisms died and were covered under layers of sand and rock. Oil deposits are most common near the Arctic, river deltas, and the edges of continents, as well as under deserts. The biggest oil deposits in the world are under the Arabian Desert in the Middle East. Oil is extracted through drilling. A rig drills a hole into the ground and pumps the oil out.

Pipelines carrying oil from deserts can leak, polluting the water supply. Oil companies are also exploring deserts to find new sources of fossil fuels. In doing so, they sometimes redefine the boundaries of protected land. For example, in 2019, the government of the West African country Niger said that it would remove 17,300 square miles (44,800 sq km) of the Sahara Desert from its protected lands.[9] This decision came after a Chinese petroleum company offered the country billions of dollars to expand its search for oil there. The move drew objections from several wildlife activists. They pointed out that the oil rigs would be located in the habitat of two endangered mammal species.

The government of Niger responded to criticisms of its decision by redrawing the boundaries of the nature preserve, extending it westward to keep the amount of protected land the same. However, this meant that the natural habitat of an antelope called the addax was no longer protected. Most of the natural land of another mammal, the dama gazelle, fell outside the new boundaries too. In the early 2020s, a group of NGOs, including the International Union for Conservation of Nature (IUCN), began working with Niger's government to find strategies for protecting these species and some of the land they live on.

Tourism and Savanna Management

The savanna grasslands of central and southern Africa are home to some of the most famous animal species on Earth. African elephants, cheetahs, giraffes, lions, and zebras are all savanna animals. Many people go on trips called safaris in order to see these animals in the wild. The desire to see animals on safari has helped motivate savanna preservation efforts. African countries such as Namibia, Zambia, and Tanzania preserve very large percentages of their land. Wildlife tourism helps these countries fund preservation efforts. However, some aspects of tourism, such as vehicle traffic, can disrupt or harm wildlife and land.

PRESERVING DESERTS AND DRYLANDS

Dryland replanting efforts and sustainable agriculture methods are two strategies people are implementing to prevent desertification before it starts. But once desertification has occurred in an area, scientists and engineers have to use different strategies to address the problem. One solution involves creating physical structures that can help keep water inside soil. For example, placing rock dams in wetland streams can help gather water from rainfall into small pools or ponds. This can then prevent soil erosion and promote the growth of plants in an area.

Tree planting is another strategy to fight desertification. Trees anchor soil, helping to prevent soil erosion. When fallen leaves decompose, they create a layer of what is known as leaf mold. Leaf mold helps the surrounding soil retain 50 percent more water than it does without the leaf layer.[10] In the drylands near the

In 2023, people in China held a workshop to introduce technology that could help countries affected by desertification control the problem.

Desert Seed Banks

One way to promote a desert's natural plant diversity is by keeping seed banks. Seed banks are collections of wild seeds representing the range of plants inside an ecosystem. For example, the Mojave Desert Land Trust (MDLT) is an organization that works on the preservation of North America's Mojave Desert. The organization collects and stores seeds from across the 41,000 square miles (105,000 sq km) of its lands.[12] The Mojave Desert Seed Bank has several goals. One is to protect against future biodiversity loss. Another is to have seeds available to share with other land preservation projects. The MDLT also educates people about the diversity of plants living in the desert.

Sahara Desert, many African countries have taken on an initiative called the Great Green Wall. This is a plan to plant a wide belt of trees that stretches from east to west across the African continent. Countries who participate in the Great Green Wall plant trees native to their areas.

While more than 20 African countries are working with organizations and agencies for this project, it has been difficult to coordinate. Between 2007, when the wall was first proposed, and 2023, only 18 percent of the wall was completed.[11] Some participating countries have chosen not to plant in line with the plans for the Great Green Wall and have instead planted trees in other regions. Over time, the plan for a Great Green Wall in some places has changed to focus on planting trees that benefit local landscapes, wildlife, and communities.

Some scientists argue that a wall of trees is not the best way to combat desertification. However, evidence suggests that coordinated tree-planting efforts have helped fight the spread of desertification. In northeastern Cameroon, the planting of 360,000 trees has helped a community of refugees obtain the shade, fertilizer, and water needed to begin growing crops in the area, which had previously been affected by desertification.[13]

In degraded deserts, some scientists are leading desert regeneration efforts that focus on restoring biodiversity. One effort involves reintroducing the biocrust. Scientists at Arizona State University have found a way to grow new desert biocrusts in a solar lab. They hope this technology can quickly restore soil in degraded deserts.

REWILDING AND WILDLIFE REINTRODUCTION

One way to try to restore dryland ecosystems, as well as other land habitats, is to leave them alone. Rewilding is the movement to restore land by not interfering with it. The idea behind rewilding is to create interconnected pieces of preserved land that give the land the resources it needs to create a thriving wild ecosystem. Rewilding started with a focus on three Cs: cores, corridors, and carnivores. *Cores* refers to protecting the core areas of preserved land. *Corridors* means establishing areas for wildlife to travel and interact. And *carnivores* refers to recovering the populations of

> **“Rewilding, in essence, is giving the land back to wildlife, and wildlife back to the land.[14]”**
>
> *—John Davis, executive director of the Rewilding Institute, 2018*

meat-eating predators. Over time, the model added three more *C*s. Those are climate resilience, compassion, and coexistence.

Rewilding seeks to minimize the effects of human influence on land. But it can include wildlife reintroduction as well. This is when native species, often predators, are brought back to a former habitat after leaving or being driven out. Rewilding can be practiced on any type of land.

One rewilding project is in the prairie grassland of the American Midwest. American Prairie is an organization that aims to restore sections of US prairies to the way they were before European settlement. The organization works in an area of Montana and seeks to acquire more land to preserve and open to the public. It removes fences on its land, making it easier for animals to travel in the area.

One of American Prairie's goals is to restore native wildlife to the prairie. American bison were once one of the most important animal species in US prairies. But during the late 1800s, bison almost disappeared

American Prairies and Preservation

The prairie grasslands of North America used to stretch north to south from Canada to Texas and from east to west from the Mississippi River to the Rocky Mountains. They covered about 266,000 square miles (689,000 sq km), an area almost the size of Texas.[15] When Europeans began settling in the prairie, they converted almost all of it into farmland and settlements. The soil of prairie grasslands was good for planting crops such as wheat and corn. Today, about 4 percent of the original tallgrass prairie is still wild land.[16] Oklahoma and Kansas are both home to tallgrass prairie preserves.

In 2009, the Nature Conservancy and the National Park Service reintroduced American bison into the Tallgrass Prairie National Preserve in Kansas to work toward restoring the prairie.

due to white settlers overhunting them. While bison have somewhat recovered from their near extinction, they are still listed as near threatened by the IUCN. American Prairie therefore manages a bison herd to help grow the species' population. For the time being, these bison are treated as semi-tame livestock, but the goal is to cultivate a wild herd of 5,000 animals on the organization's land.[17]

A longer-term goal for American Prairie is to reintroduce large predators to the prairies, including grizzly bears and wolves. These animals are naturally returning to wild land. Rewilding organizations keep an eye on this process, making sure that the return is as safe as possible for both the humans in the area and the animals. American Prairie also works to reseed native prairie grasses. They hope that this reseeding will naturally attract a greater variety of animals, such as insects and birds.

In the winter, the average temperature in the Arctic is about –30 degrees Fahrenheit (–34°C), and in Antarctica it's about –76 degrees Fahrenheit (–60°C).

CHAPTER 7

ARCTIC AND ANTARCTIC PRESERVATION

The region near the North Pole is called the Arctic, and the southern polar region is known as Antarctica. These regions face unique issues in preserving land. The Arctic and Antarctica are greatly affected by climate change. Global warming is causing glaciers in these regions to melt at a quicker rate than usual. This melting process can change the surrounding land.

When glaciers melt, they release excess water and send nutrients into surrounding water sources. Land that was once underneath the ice then becomes exposed to the environment. This exposure can have wide-ranging effects on Arctic and Antarctic biomes.

The Arctic and Antarctica are also politically complex places. Arctic land belongs to several different countries, while Antarctica belongs to none. Many different entities and organizations have interests in exploring or using Arctic and Antarctic land. But international communities disagree about how these regions should be used or preserved.

MELTING GLACIERS AND LAND PRESERVATION

Glaciers form when heavy snowfall stays on the ground, eventually compacting into ice. Throughout the planet's history, Earth's average temperature has risen and fallen.

Land Preservation and the North Pole

The land around the North Pole is technically sea ice. The ice cover grows thicker during winter months and shrinks to its lowest point in September. This ice connects to Greenland, Russia, and Canada. Scientists began measuring the extent of Arctic sea ice with satellites in 1979. In September 1980, Arctic sea ice covered more than 2.9 million square miles (7.5 million sq km). In the same month in 2012, it covered only 1.3 million square miles (3.4 million sq km). By September 2023, however, measurements showed that the sea ice covered up to 1.7 million square miles (4.4 million sq km).[2]

Low-temperature periods make glaciation, the forming of glaciers, more likely to happen. Eras with a lot of glaciation are called ice ages.

When glaciers melt, they change the landscape around them. Ice and meltwater help create landforms such as valleys by eroding soil and rock. Today, about 10 percent of the planet's surface is covered by glaciers.[1] Antarctica has the most glacial ice in the world, followed by the Arctic region of Canada. Alaska also has significant glacial coverage, although much of it is in the subarctic region, the area just below the Arctic Circle.

Global temperatures are rising due to climate change. As a result, glaciers are melting, in many cases rapidly. Glacial melting

Pieces of ice more than 50 miles (80.5 km) long can break off the ice shelves in Antarctica.

has many effects on the land that was previously covered by glaciers, as well as the land and waterways around it. When glacier ice melts, the meltwater drains into the ocean, leading to rising sea levels. Rising sea levels, in turn, erode coastal land elsewhere.

Glacier and polar ice melting tend to accelerate over time. This is because of a process known as ice-albedo feedback. Ice is

Emperor penguin populations in some areas of Antarctica have declined up to 50 percent due to melting sea ice.

much better at reflecting the sun's heat than the land and ocean water are. As the amount of ice on Earth's surface shrinks, the area around the ice takes in more heat from the sun than before. Thus, a decrease in ice cover leads to an increase in temperature, which in turn contributes to the melting of more ice.

Melting glaciers expose the land underneath the ice to air and sunlight. In this way, glacier melting makes more land available to plants, animals, and other wildlife. Scientists predict that between 2000 and 2100, the amount of land covered by glaciers may reduce by more than 20 percent.[3] They aren't sure how wildlife will begin to live in this changed land. In addition, scientists don't know how these new ecosystems will affect the surrounding land.

Since 1993, meltwater from ice sheets in Antarctica and Greenland have contributed to about one-third of the average global sea level rise.[5]

Glacial melting also opens land to human use. There are large oil and natural gas deposits in the land underneath and near glaciers. There are also fossil fuel deposits underneath the ice that partially covers the Arctic Ocean. These fossil fuel deposits are the property of the countries in which they're located, such as Russia, Canada, and the United States.

Increased extraction of these reserves comes with environmental risks. For example, when oil companies extract oil from the Arctic, they practice something called gas flaring, or burning a gas released during extraction. This practice greatly increases carbon emissions in the region and contributes to global warming.

CLIMATE CHANGE, PERMAFROST, AND ARCTIC LAND

In very cold areas of the world, the land both near and underneath glaciers has a permanently frozen layer. This layer of soil or rock is called permafrost. Permafrost is also an important carbon sink. About half of the carbon stored in the world's soil is stored in permafrost.[4] Most of the world's permafrost is in the

Arctic region, and the biome that supports plant, animal, and other life there is called the tundra.

Tundra life-forms are uniquely adapted to withstand very cold temperatures and frozen soil. For example, lichens are combinations of fungi and bacteria that provide food sources for animals in places where the soil is frozen and plants are scarce. As temperatures rise, however, permafrost layers begin melting. Permafrost is heating faster than Arctic air temperatures. Between 1990 and 2020, permafrost temperatures rose between 2.7 and 4.5 degrees Fahrenheit (1.5–2.5°C).[6]

As Earth's climate is heating up, permafrost is melting. Lichens adapt slowly to climate change. Experts believe lichens could take more than a million years to adjust to a global temperature increase of 1.8 degrees Fahrenheit (1°C).[7] As permafrost melts, lichens may begin to disappear. This disruption of the ecosystem will ripple out to the other wildlife in the tundra. Caribou are a deerlike animal that live in the tundra and migrate from place to place in herds. Lichens are their main food source. As lichen coverage decreases and plants replace this life-form, caribou populations in the tundra have declined.

> **"It's a big question mark as to what Antarctica will look like in a few decades if we don't reduce our carbon emissions.[8]"**
>
> *—Claire Christian, director of the Antarctic and Southern Ocean Coalition, 2022*

Melting permafrost can also damage human-made structures. As the ice inside permafrost melts, the ground becomes wet, causing buildings to sink. In addition, areas of land covered by glaciers and permafrost have historically been used as toxic waste dumping grounds. Because these areas were frozen or covered, people believed the materials left there would stay contained. But as glaciers and permafrost melt, these toxic materials, including radioactive waste, may be rereleased into the environment.

ANTARCTICA, CLIMATE CHANGE, AND LAND PRESERVATION

Antarctica is a unique continent. It is the coldest, driest, and windiest continent in the world, and 98 percent of its land is covered by ice.[9] Unlike the world's other landmasses, Antarctica has no countries. Instead, it is subject to a joint international agreement and regulated by a document called the Antarctic Treaty. No one lives there permanently, and there are many rules about who can visit the land. Tourists who visit Antarctica are not allowed to litter, bother wildlife, or take back artifacts such as rocks.

Although laws and regulations are in place to protect Antarctic land, they are not always followed. The continent faces threats from invasive species brought by tourists. For example, non-native mites and earthworms have begun living on the continent. In addition, people who visit Antarctica can carry extremely small organisms with them, including strains of bacteria. These bacteria can infect wildlife, bringing new diseases into the region. There are records of bacteria traveling

Tourists can take boat rides around parts of Antarctica to view wildlife.

from humans to Antarctic penguins, which could affect penguin populations.

The ice that covers Antarctica is not melting as fast as Arctic ice. This is because Antarctica is at a higher elevation, its ice covers a larger surface, and the Southern Ocean helps cool the continent. However, the continent is still losing about 150 billion short tons (136 billion metric tons) of ice per year.[10] Scientists predict that as Earth continues to warm, Antarctic ice will begin melting even more rapidly.

The Antarctic Treaty states that Antarctic land should be used for peaceful purposes, including scientific research. Commercial activities such as mining are not allowed. Bioprospecting, or the search for living things such as microbes for human use, does occur in Antarctica. Meanwhile, some countries and companies are interested in finding and using the minerals and fossil fuels that are underneath Antarctic ice. In 2020, the Russian government released a statement saying that it believed that there were up to 500 billion barrels of fossil fuel resources under Antarctic ice.[11] Many believed this statement was a way of showing Russia's interest in extracting resources on the continent.

PRESERVING ARCTIC AND ANTARCTIC LAND

The fight to preserve Arctic and Antarctic land is connected with the fight against global warming. Researchers, policymakers, and organizations are working together to try to slow the rate of ice melting at the poles. For example, the Arctic Council is a group of eight countries and six Indigenous councils with land in the Arctic. The council is joined by 38 observers, which include non-Arctic states and NGOs.

The Arctic Council works to address the global threat of climate change while also focusing on environmental issues that uniquely affect the Arctic region. For example, the council studies the effect of black

Bioprospecting

Bioprospecting is the search for living things that can be used in human goods. Bioprospectors look for sources of new medicines or products in plants and other life-forms. Some bioprospectors explore Antarctic land. They have found organisms such as a type of yeast that can help prevent tumors. However, the use of Antarctic land for bioprospecting is controversial. Some people believe that the threat to wildland outweighs the potential benefits.

Eklutna

Eklutna is a village in Alaska. Its name comes from the Dena'ina language and refers to the two pieces of granite near a local river. In 1971, a law called the Alaska Native Claims Settlement Act gave money and land to Native Alaskan people in exchange for eliminating Indigenous peoples' rights to Alaskan land. The following year, a group of Alaska Natives came together to create Eklutna Inc., a business that works to maintain cultural practices and economic stability in Eklutna. In 2014, the business bought land and created the first Indigenous conservation easement. The easement allows for Indigenous peoples to continue traditional land use in the area.

carbon, the air pollutant released from gas flaring in Arctic oil drilling. It submits reports monitoring each country's efforts to reduce this pollutant and makes recommendations for further progress that can be made.

The Indigenous governing bodies on the council also play a big role in Arctic land preservation. Across the Arctic, Indigenous communities have come together to lead efforts to set aside land for sustainable use. In 2018, the Inuit people of Greenland cooperated with the UN to set aside land in the country as a cultural heritage site. Within this site, mining and resource extraction are banned, while sustainable hunting and fishing practices continue.

In the United States, many conflicts over Arctic land preservation center on protecting land from the consequences of oil and gas drilling. In September 2023, Secretary of the Interior Deb Haaland approved an official cancellation for the planned leases of land in the US Arctic Wildlife Refuge. This land was set to be rented to oil and gas companies until the government started intervening to protect the land in 2021.

In addition, the US government created new regulations for Alaska's National Petroleum Reserve. This area covers 36,000 square miles (93,000 sq km) of land in northern Alaska and was originally set aside as a source of oil for the US Navy. About half of this land is preserved for wildlife and for use by Alaskan Native groups. The other half is open for drilling.[12]

While new regulations require companies to drill with maximum protection of the environment in the reserve, it is not clear what that will mean in practice. In 2023, the US government approved a large oil drilling project on the lands. Some environmental activists believe that the government should cancel more oil and gas leases in the region.

Researchers gather data about Antarctica and the Arctic to understand the effects of climate change on the polar land and sea.

When visiting national or state parks, it's important for people to follow park rules to prevent damaging the environment.

CHAPTER 8

GETTING INVOLVED IN LAND PRESERVATION

Land preservation is an important global issue, but to some it can seem like a difficult problem to tackle. The land on Earth is limited, and not all changes to land are reversible. Furthermore, it can be hard for people to preserve land on their own. Decisions about land preservation are often in the hands of wealthy individuals, private companies, and governments.

However, there is a great deal that ordinary people can do to join land preservation efforts. People can use their own time and skills to help clean and maintain local land. They can educate themselves and others on the importance of land preservation. And they can influence decisionmakers to take action to preserve land. Everyday people worldwide are taking important steps to preserve land for themselves and for future life on Earth.

EVERYDAY EFFORTS

Many land preservation efforts are local in scale. Preserved land must be managed so that it stays in good condition. But it can sometimes be difficult to ensure that people are following the rules set in preserved land areas. For example, national parks do not allow people to get close to wildlife. However, many people break these rules, attempting to take pictures with animals such as elk or bison.

Litter and trash are also common issues on public land. Those who maintain the land will often organize cleanup days or projects. At these cleanups, volunteers work as land managers, helping to clear litter and trash. Sometimes, young people can earn community service hours for doing cleanup work in their communities.

People of all ages who want to help clean preserved land can go online and find opportunities to volunteer. Volunteer cleanup opportunities can be found nearly anywhere. More than 220,000 square miles (570,000 sq km) of US forests are located inside cities and towns.[1] Volunteers are needed to help keep these forests clean and healthy. Private land trusts offer similar volunteer opportunities. The Land Trust Alliance allows people to search for opportunities to volunteer at trusts near them.

> **"It's surely our responsibility to do everything within our power to create a planet that provides a home not just for us, but for all life on Earth.[2]"**
>
> ***—David Attenborough, biologist,* Planet Earth II, *2016***

The National Park Service also offers paid jobs through its Youth Conservation Corps. These are eight-to-ten-week opportunities that teens ages 15 to 18 can take part in. As part of the conservation corps, teens can learn how to do land management work such as cleaning campgrounds and preserving historical sites. The corps at Yellowstone offers a residential program as well, which means that workers can stay at the camp during their time there.

People can also work to fight deforestation by joining tree-planting efforts. In 1970, Andy Lipkis was a 15-year-old camp counselor who was concerned about the deforestation of California forests. He founded TreePeople to inspire others to plant trees. TreePeople is a tree-planting organization based in Southern California. It was founded in 1973 and was originally called the California Conservation Project. The organization offers weekly volunteer opportunities for people who want hands-on experience planting trees and reforesting land.

Litter Cleanups

One way people work to preserve land is by picking up litter, or discarded trash that is not put in a waste container. Litter cleanups are organized efforts to gather this litter and dispose of it properly. The first anti-litter campaigns in the United States started in the 1950s. Some of the first organized beach cleanups took place in the mid-1980s. And in 1994, the National Environmental Education Foundation organized the first National Public Lands Day. This was a day urging people to volunteer to clean up local public spaces. Since then, the organization and the National Park Service have helped publicize volunteer cleanup opportunities.

Evidence suggests that land preservation work can be good for people's health. New York's Department of Environmental Conservation found that being around trees or in forests can help boost people's immune system function and mood. Being in nature has also been shown to reduce stress levels.

These health benefits may be especially strong for children and teens. A 2020 study published in the *International Journal of Environmental Research and Public Health* found that young people who had access to green spaces such as parks were likely to have improved mental health. This was true for a variety of mental health issues, from depression to symptoms of attention deficit hyperactivity disorder (ADHD).

People can help restore ecosystems by volunteering to remove non-native plants.

PROMOTING LAND PRESERVATION

Everyday citizens can also play a part in organizing land preservation efforts. They can educate others about land issues and get the word out about important land preservation battles. Many NGOs run multiple campaigns at the same time. They try to draw attention and funding to certain issues and urge government leaders to take action. These NGOs often rely on volunteers to help with this publicity work.

For example, Greenpeace is an organization that brings together environmental issues from around the world. The organization's website collects a list of ongoing land preservation initiatives. One initiative is the fight to end the clearing of Amazon rainforest.

Net-Zero Carbon and Tree Planting

Some private companies make a pledge to have net-zero carbon emissions. This is when they balance the carbon emissions from activities such as transporting goods or running factories with steps to absorb carbon. One strategy for achieving net-zero emissions is planting trees. While all trees absorb carbon, experts warn that planting the same species of tree can have negative effects. For example, many companies sponsor the planting of eucalyptus trees. But planting forests with one type of tree does not support biodiversity as well as natural forests do. Also, because trees take time to grow, planting trees does not immediately offset carbon emissions.

Land preservation advocacy can also involve contacting government officials to get more regulations put in place. Government officials are responsible to the people who elect them to office. When people contact government officials about important issues, it shows the officials that people believe the issues need to be addressed. This information can affect the way officials vote on different laws and regulations, including laws to reserve land or devote more funding to conservation projects.

Other land preservation organizations use similar tactics. The Nature Conservancy is a nonprofit organization that promotes land and water conservation. The organization asks volunteers to help out by signing open letters to send to members of Congress.

One Nature Conservancy campaign asks the government to stick to its goal of preserving 30 percent of US land and water by 2030.[3] The organization's website offers a place for volunteers to sign this letter. Its goal is to show members of Congress that the public is serious about holding the government accountable on environmental issues.

People can also get involved in land preservation by educating others about the issue. Some organizations allow people to get involved in the administration of a piece of public land. Organizations such as the Natural Land Institute also allow volunteers to work on community outreach and engagement related to this issue. Volunteers help host events that teach people about the importance of land preservation.

TEENS MAKING A DIFFERENCE

There are many inspiring teens making a difference in the world of land preservation. Some are becoming government representatives, working with leaders to set policy. In 2022, 16-year-old Logan Sizemore became the youngest person to win a government election in Kentucky. Sizemore ran for and won a seat on his local Soil and Water Conservation District board.

As a member of the board, Sizemore helps decide how to preserve and beautify local land. He is also interested in agriculture and educating other teens about agricultural methods. "I want to make someone come look at all parts of our county and say this is one of the most beautiful places in all of Kentucky," Sizemore said.[4]

For the teens of the Tribal Youth Conservation Corps in Texas, improving the look of land is a hands-on project. The Tribal Youth Conservation Corps is a program that offers American Indian teens opportunities to get involved with conservation. Members of the program learn advanced land management skills such as fire management. They also learn about wildlife significant to their cultural heritage. Members of the corps can get hands-on

Letter-Writing Campaigns

Anyone under the age of 18 is ineligible to vote, and most teens cannot make financial decisions such as buying conservation easements. But teens can still take part in more involved land preservation efforts. One way they can get involved is through a letter-writing campaign. Letter-writing campaigns are designed to get the attention of elected officials and others in power. They can also help media outlets decide if an issue is worth covering. In addition, these campaigns can help raise funds for important issues such as land preservation.

ACTIVIST SPOTLIGHT

HELENA GUALINGA

Sumak Helena Sirén Gualinga is a speaker and activist from Ecuador. She was born in a small forest community that is home to the Kichwa Sarayaku people. Gualinga's mother, Noemi, is a leader in the community.

The Sarayaku peoples' land came under threat in 2002 when the oil company Chevron and the Ecuadorian government put explosives in the rainforest and dug test wells to try to find resources underground. These activities ultimately damaged the land. In 2012, the Sarayaku people fought the Ecuadorian government in an international human rights court and won.

Gualinga was inspired by her family and peoples' human rights work. She began speaking about the destruction of the Amazon rainforest in local schools. Gualinga helped form an organization called Polluters Out. The organization draws attention to the fact that energy companies such as Chevron influence climate change decision making.

Gualinga's work has been featured in a documentary and the magazine *Vogue*. Gualinga sees land preservation, fighting climate change, and human rights activism as related to each other. "I believe in change and that humanity is not separated from nature," she explains.[5] She believes that as a woman and young Indigenous person, she brings an important perspective to international conversations about climate change.

Along with being an environmental activist, Helena Gualinga is an activist for Indigenous peoples' rights.

experience and decide whether they want to pursue a career in land management.

In New Mexico, teens from the Indigenous land in Jemez Pueblo pursue similar work with the Forest Stewards Youth Corps. Owen Pecos is a volunteer with the program. He said, "I just want to help my community for the next generation."[6]

In 2022, Secretary of the Interior Deb Haaland announced a nationwide program called the Indian Youth Service Corps. This program helps empower Indigenous youth to manage and preserve traditional land. The program offers education, training, and work experience on land management issues. Participants can also work on projects such as trail restorations and help create educational materials for visitors about the preserved land.

A 2019 survey found that 86 percent of US voters supported conservation efforts to protect 30 percent of land and water by 2030.[7]

Around the world, people are making their voices heard to preserve important and vulnerable lands. Together, volunteers, advocates, educators, and organizations are working to protect land for the good of the planet and for future generations. Anyone can take action to help preserve and restore the land on Earth.

ESSENTIAL FACTS

LAND PRESERVATION PROBLEMS

- Human activities such as development and fossil fuel extraction can degrade land and damage ecosystems.
- Deforestation and agricultural practices, especially in the Amazon rainforest, are threatening forests around the world. Climate change is also causing forests to be more susceptible to wildfires.
- Wetland ecosystems help purify water and are important carbon sinks. But pollution, climate change, and land development are putting these ecosystems at risk.
- Desertification is degrading the soil in drylands, making it more difficult for plants to grow in those environments.
- Climate change has increased glacial ice melting in polar regions, which changes ecosystems and causes sea levels to rise.

LAND PRESERVATION SOLUTIONS

- Combating the effects of climate change to help reduce global threats such as glacial melting and desertification.
- Protecting land from harmful human use at a state, national, or international level to ensure vulnerable habitats are not damaged even more.
- Implementing sustainable farming practices to help lessen the damage caused to land used for agricultural purposes.
- Replanting forests and rewilding biomes to reverse some of the effects of land degradation.
- Maintaining Arctic and Antarctic regions to prevent companies from exploring the land for fossil fuel and mineral deposits.

TAKING ACTION

- Volunteer to clean up garbage and to plant trees to help preserve land.
- Contact government representatives to get more land preservation laws put in place in order to protect vulnerable ecosystems.
- Educate others about the importance of land preservation to raise awareness of the issue.

QUOTE

"We owe it to ourselves and to the next generation to conserve the environment so that we can bequeath our children a sustainable world that benefits all."

—*Wangari Maathai, founder of the Green Belt Movement*

GLOSSARY

algae
Plantlike organisms that mostly grow in water and are often green, blue, red, or brown colored.

atmosphere
The layer of gases surrounding a planet.

biodiversity
The many different plants and animals in an ecosystem.

degrade
To reduce a habitat's quality.

domesticate
To adapt to live among or be of use to people.

ecosystem
A community of interacting organisms and their environment.

emission
The production and discharge of something such as smoke, gas, or chemicals into the air.

extinction
The state of having completely died out.

famine
An extreme scarcity of food.

fertilizer
A substance with ingredients that help plants grow.

glacier
A large, thickened mass of ice formed from fallen snow that has compacted and been added to over many years.

habitat
The natural environment of an organism.

irrigate
To supply water to land or crops to help with growth.

migrate
In animals, to travel relatively long distances between locations, usually to escape unfavorable conditions or to find food.

nongovernmental organization (NGO)
An organization that has a social mission and is not associated with a government.

sustainable
A practice that avoids depleting natural resources.

toxic
Poisonous to an animal or habitat.

treaty
A formal agreement between two or more countries or leaders.

watershed
A region or area in which water drains into the same body of water.

ADDITIONAL RESOURCES

SELECTED BIBLIOGRAPHY

"Earth's Biodiversity Depends on the World's Forests." *United Nations Environment Programme*, May 2020, unep-wcmc.org. Accessed 16 Jan. 2024.

Imbler, Sabrina. "Meet Peat, the Unsung Hero of Carbon Capture." *New York Times*, 22 Feb. 2022, nytimes.com. Accessed 16 Jan. 2024.

Nunez, Christina. "Desertification, Explained." *National Geographic*, 31 May 2019, nationalgeographic.com. Accessed 16 Jan. 2024.

FURTHER READINGS

Jahren, Hope. *The Story of More (Adapted for Young Adults): How We Got to Climate Change and Where to Go from Here*. Ember, 2021.

Pelleschi, Andrea. *Amazon Explorers*. Abdo, 2020.

Rowell, Rebecca. *Energy Extraction*. Abdo, 2025.

ONLINE RESOURCES

Booklinks
NONFICTION NETWORK
FREE! ONLINE NONFICTION RESOURCES

To learn more about land preservation, please visit **abdobooklinks.com** or scan this QR code. These links are routinely monitored and updated to provide the most current information available.

MORE INFORMATION

For more information on this subject, contact or visit the following organizations:

Land Trust Alliance
1250 H St. NW, Ste. 600
Washington, DC 20005
landtrustalliance.org
The Land Trust Alliance works to create and maintain private land trusts and invests in a wide variety of land preservation projects.

Nature Conservancy
4245 N Fairfax Dr., Ste. 100
Arlington, VA 22203
nature.org
The Nature Conservancy works to combat the interrelated threats of climate change and biodiversity loss. Among other initiatives, it also aims to preserve 2.5 million square miles (6.5 million sq km) of land by 2030.

World Wildlife Fund (WWF)
1250 24th St. NW
Washington, DC 20090
worldwildlife.org
The World Wildlife Fund (WWF) works to conserve endangered species by maintaining biodiversity and the sustainable use of natural resources.

SOURCE NOTES

CHAPTER 1. SAVE THE SWAMP

1. "Gulf Sturgeon." *National Oceanic and Atmospheric Administration Fisheries*, 1 Nov. 2022, fisheries.noaa.gov. Accessed 5 Mar. 2024.

2. "America's Sea Level Has Risen by 6.5 Inches since 1950." *SeaLevelRise.org,* 2016, sealevelrise.org. Accessed 5 Mar. 2024.

3. "Louisiana's Sea Level Is Rising." *SeaLevelRise.org,* 2016, sealevelrise.org. Accessed 5 Mar. 2024.

4. "2022 Sea Level Rise Technical Report." *National Oceanic and Atmospheric Administration*, 2022, oceanservice.noaa.gov. Accessed 5 Mar. 2024.

CHAPTER 2. THE SCIENCE OF LAND PRESERVATION

1. "The Earth." *Nations Online Project*, n.d., nationsonline.org. Accessed 5 Mar. 2024.

2. Elizabeth Pennisi. "Just 19% of Earth's Land Is Still 'Wild,' Analysis Suggests." *Science*, 19 Apr. 2021, science.org. Accessed 5 Mar. 2024.

3. Drew McConville and Angelo Villagomez. "Why Conserving 30x30 Is More Than a Numbers Game." *Center for American Progress*, 31 Jan. 2023, americanprogress.org. Accessed 5 Mar. 2024.

4. Terri Lane. "Conservation vs. Preservation." *Northwest Arkansas Land Trust*, 10 Aug. 2020, nwalandtrust.org. Accessed 5 Mar. 2024.

5. "Lower Gallatin Watershed Resources." *Gallatin Watershed Council*, n.d., gallatinwatershedcouncil.org. Accessed 5 Mar. 2024.

6. "The Ocean—The World's Greatest Ally against Climate Change." *United Nations*, n.d., un.org. Accessed 5 Mar. 2024.

7. "What Is Climate Change?" *United Nations*, n.d., un.org. Accessed 5 Mar. 2024.

8. "Carbon Dioxide." *National Aeronautics and Space Administration*, n.d., climate.nasa.gov. Accessed 5 Mar. 2024.

9. "Rare Plant Conservation." *Bureau of Land Management*, n.d., blm.gov. Accessed 5 Mar. 2024.

10. "World Met Target for Protected Area Coverage on Land, but Quality Must Improve." *United Nations Environment Programme*, 19 May 2021, unep.org. Accessed 5 Mar. 2024.

CHAPTER 3. THE HISTORY OF LAND PRESERVATION

1. Erle C. Ellis, et al. "People Have Shaped Most of Terrestrial Nature for at Least 12,000 Years." *Proceedings of the National Academy of Science*, 19 Apr. 2021, pnas.org. Accessed 5 Mar. 2024.

2. Hannah Ritchie and Max Roser. "Half of the World's Habitable Land Is Used for Agriculture." *Our World in Data*, 16 Feb. 2024, ourworldindata.org. Accessed 5 Mar. 2024.

3. Krishan Kumar. "Population Change." *Britannica*, n.d., britannica.com. Accessed 5 Mar. 2024.

4. Erle C. Ellis, et al. "Anthropogenic Transformation of the Biomes, 1700 to 2000." *Global Ecology and Biogeography*, 4 Aug. 2010, onlinelibrary.wiley.com. Accessed 5 Mar. 2024.

5. "Theodore Roosevelt Quotes." *National Park Service*, 10 Aug. 2015, nps.gov. Accessed 5 Mar. 2024.

6. "From Gold Rush to Golden State." *Library of Congress*, n.d., loc.gov. Accessed 5 Mar. 2024.

7. "Goal 15: Life on Land." *Global Partnership for Sustainable Development Data*, n.d., data4sdgs.org. Accessed 5 Mar. 2024.

8. "Monuments, Conservation Areas and Similar Designations." *Bureau of Land Management*, n.d., blm.gov. Accessed 5 Mar. 2024.

9. Rocío Lower and Rebecca Watson. "How Many National Parks Are There?" *National Park Foundation*, 11 Dec. 2023, nationalparks.org. Accessed 5 Mar. 2024.

10. "The Everglades." *National Wildlife Federation*, n.d., nwf.org. Accessed 5 Mar. 2024.

CHAPTER 4. FOREST PRESERVATION

1. Michael Kuhns. "What Is a Tree?" *Utah State University Forestry Extension*, n.d., extension.usu.edu. Accessed 5 Mar. 2024.

2. "Boreal Forest/Taiga Biome." *Let's Talk Science*, 3 Feb. 2020, letstalkscience.ca. Accessed 5 Mar. 2024.

3. Michael Carlowicz. "Seeing Forests for the Trees and the Carbon: Mapping the World's Forests in Three Dimensions." *National Aeronautics and Space Administration Earth Observatory*, 9 Jan. 2012, earthobservatory.nasa.gov. Accessed 5 Mar. 2024.

4. "Forests and the Global Biodiversity Framework: Expectations and Must-Haves." *World Wildlife Fund*, 1 Dec. 2022, wwf.panda.org. Accessed 5 Mar. 2024.

5. Carlowicz, "Seeing Forests for the Trees."

6. Christina Nunez. "Why Deforestation Matters—And What We Can Do to Stop It." *National Geographic*, 7 Dec. 2022, nationalgeographic.com. Accessed 5 Mar. 2024.

7. "What Animals Live in the Amazon? And 8 Other Amazon Facts." *World Wildlife Fund*, n.d., worldwildlife.org. Accessed 5 Mar. 2024.

8. "Amazon Facts." *World Wildlife Fund*, n.d., worldwildlife.org. Accessed 5 Mar. 2024.

9. "Top Facts about the Amazon." *World Wildlife Fund United Kingdom*, 21 Aug. 2023, wwf.org.uk. Accessed 5 Mar. 2024.

10. Grace Hussain. "Amazon Deforestation—How Much of the Rainforest Is Left?" *Sentient Media*, 13 Oct. 2023, sentientmedia.org. Accessed 5 Mar. 2024.

11. Carlowicz, "Seeing Forests for the Trees."

12. "Deforestation & Palm Oil." *Sustainable Palm Oil Choice*, n.d., sustainablepalmoilchoice.eu. Accessed 5 Mar. 2024.

13. "Zoonotic Diseases." *Centers for Disease Control and Prevention*, 1 July 2021, cdc.gov. Accessed 5 Mar. 2024.

14. "Amazon Rainforest." *Britannica*, 29 Feb. 2024, britannica.com. Accessed 5 Mar. 2024.

15. Paul Brinkmann. "Florida Professor Predicts Amazon Rainforest Collapse by 2064." *United Press International*, 29 Dec. 2020, upi.com. Accessed 5 Mar. 2024.

16. Charles, Prince of Wales. "HRH the Prince of Wales Delivers the Presidential Lecture, Presidential Palace, Jakarta, Indonesia." *Royal Household*, 3 Nov. 2008, royal.uk. Accessed 5 Mar. 2024.

17. "Recognizing Indigenous Peoples' Land Interests Is Critical for People and Nature." *World Wildlife Fund*, 22 Oct. 2020, worldwildlife.org. Accessed 5 Mar. 2024.

18. Nadia S. Santini and Yosune Miquelajauregui. "The Restoration of Degraded Lands by Local Communities and Indigenous Peoples." *Frontiers*, 25 Apr. 2022, frontiersin.org. Accessed 5 Mar. 2024.

19. "Ghana Receives Its First Payment for Emissions Reductions." *United Nations Environment Programme*, 5 Apr. 2023, unep.org. Accessed 5 Mar. 2024.

20. "Ghana Receives Its First Payment."

21. "We Are One Tree Planted: Our Vision." *One Tree Planted*, 2024, onetreeplanted.org. Accessed 5 Mar. 2024.

22. "Kenya: Summary." *Global Forest Watch*, n.d., globalforestwatch.org. Accessed 5 Mar. 2024.

23. Jason Daniel Nuckolls. "Kenyan Footballer Lesein Mutunkei Scores Environmental Goals with Tree-Planting Project." *Newsweek*, 8 Sept. 2023, newsweek.com. Accessed 5 Mar. 2024.

24. "Lesein Mutunkei." *It's Time*, n.d., itstime.earth. Accessed 5 Mar. 2024.

SOURCE NOTES CONTINUED

CHAPTER 5. WETLAND PRESERVATION

1. Thomas E. Dahl and Gregory J. Allord. "History of Wetlands in the Conterminous United States." *US Geological Survey*, n.d., water.usgs.gov. Accessed 5 Mar. 2024.

2. "Why Wetlands?" *Wildfowl and Wetlands Trust*, wwt.org.uk. Accessed 5 Mar. 2024.

3. "Recommendations for Reducing Wetland Loss in Coastal Watersheds of the United States." *National Oceanic and Atmospheric Administration Fisheries*, 13 June 2022, fisheries.noaa.gov. Accessed 5 Mar. 2024.

4. "'Wetlands Are Being Lost Faster than Forests.'" *Ramsar*, 13 June 2017, ramsar.org. Accessed 5 Mar. 2024.

5. "How Much of the Earth's Water Is Stored in Glaciers?" *US Geological Survey*, n.d., usgs.gov. Accessed 5 Mar. 2024.

6. "'Wetlands Are Being Lost.'"

7. "Peatlands." *Department of Agriculture, Environment, and Rural Affairs*, n.d., daera-ni.gov.uk. Accessed 5 Mar. 2024.

8. Sabrina Imbler. "Meet Peat, the Unsung Hero of Carbon Capture." *New York Times*, 22 Feb. 2022, nytimes.com. Accessed 5 Mar. 2024.

9. "Partnerships with Landowners under CWA Section 404." *Environmental Protection Agency*, 11 Jan. 2024, epa.gov. Accessed 5 Mar. 2024.

10. "Drive to Protect World's Wetlands Gains Momentum." *United Nations Environment Programme*, 1 Feb. 2023, unep.org. Accessed 5 Mar. 2024.

CHAPTER 6. DRYLAND PRESERVATION

1. Christina Nunez. "Deserts Explained." *National Geographic*, 19 Oct. 2023, education.nationalgeographic.org. Accessed 5 Mar. 2024.

2. Peter Laban. "IUCN CEM Dryland Ecosystems Specialist Group." *International Union for Conservation of Nature*, n.d., iucn.org. Accessed 5 Mar. 2024.

3. Christina Nunez. "Desertification, Explained." *National Geographic*, 31 May 2019, nationalgeographic.com. Accessed 5 Mar. 2024.

4. Sofie Bates. "NASA-Funded Scientists Estimate Carbon Stored in African Dryland Trees." *National Aeronautics and Space Administration*, 1 Mar. 2023, nasa.gov. Accessed 5 Mar. 2024.

5. "Top Facts about the Amazon." *World Wildlife Fund United Kingdom*, 21 Aug. 2023, wwf.org.uk. Accessed 5 Mar. 2024.

6. Niall P. Hanan, et al. "A Role for Drylands in a Carbon Neutral World?" *Frontiers*, 5 Nov. 2021, frontiersin.org. Accessed 5 Mar. 2024.

7. Stuart L. Pimm and John P. Rafferty. "Desertification." *Britannica*, 18 Jan. 2024, britannica.com. Accessed 5 Mar. 2024.

8. "High-Level Dialogue on Desertification, Land Degradation and Drought." *United Nations*, 14 June 2021, un.org. Accessed 5 Mar. 2024.

9. Stephen Leahy. "Expanding Oil Exploration Threatens One of the Sahara's Largest Nature Reserves." *Pacific Standard*, 2 Aug. 2019, psmag.com. Accessed 5 Mar. 2024.

10. Paul Barbano. "Leaf Mold Benefits Soil Structure and Water Retention." *Cape Gazette*, 18 Nov. 2020, capegazette.com. Accessed 5 Mar. 2024.

11. Spoorthy Raman. "Progress Is Slow on Africa's Great Green Wall, but Some Bright Spots Bloom." *Mongabay*, 3 Aug. 2023, news.mongabay.com. Accessed 5 Mar. 2024.

12. "The Mojave Desert Seed Bank." *Mojave Desert Land Trust*, n.d., mdlt.org. Accessed 5 Mar. 2024.

13. Sharon Newell. "Five Ways the Great Green Wall Is Helping to Reforest the Sahel Region." *BBC Earth*, n.d., bbcearth.com. Accessed 5 Mar. 2024.

14. John Davis. "Rewilding Distilled." *Rewilding Institute*, 21 July 2018, rewilding.org. Accessed 5 Mar. 2024.

15. "Five Facts about America's Once Vast Prairie." *Forest Preserve District of Will County*, 27 Aug. 2021, reconnectwithnature.org. Accessed 5 Mar. 2024.

16. "A Complex Prairie Ecosystem." *National Park Service*, n.d., nps.gov. Accessed 5 Mar. 2024.

17. "Bison Restoration." *American Prairie*, n.d., americanprairie.org. Accessed 5 Mar. 2024.

CHAPTER 7. ARCTIC AND ANTARCTIC PRESERVATION

1. "Quick Facts: What Is a Glacier?" *National Snow and Ice Data Center*, n.d., nsidc.org. Accessed 15 Mar. 2024.

2. "Arctic Sea Ice Minimum Extent." *National Aeronautics and Space Administration*, n.d., climate.nasa.gov. Accessed 5 Mar. 2024.

3. Jeff Tollefson. "Melting Glaciers Will Reveal Vast New Ecosystems in Need of Protection." *Nature*, 16 Aug. 2023, nature.com. Accessed 15 Mar. 2024.

4. Alina Bykova. "Permafrost Thaw in a Warming World: The Arctic Institute's Permafrost Series Fall-Winter 2020." *Arctic Institute*, 1 Oct. 2020, thearcticinstitute.org. Accessed 5 Mar. 2024.

5. "Ice Sheets." *National Aeronautics and Space Administration*, n.d., climate.nasa.gov. Accessed 5 Mar. 2024.

6. Bykova, "Permafrost Thaw in a Warming World."

7. Corryn Wetzel. "Lichens Are in an Evolutionary Race against Climate Change." *Smithsonian Magazine*, 17 Feb. 2022, smithsonianmag.com. Accessed 5 Mar. 2024.

8. Tara Lohan. "Action for Antarctica: Saving the World's Last Great Wilderness." *Revelator*, 17 Aug. 2022, therevelator.org. Accessed 5 Mar. 2024.

9. Lize-Marié van der Watt. "Antarctica." *Britannica*, 3 Mar. 2024, britannica.com. Accessed 5 Mar. 2024.

10. "Ice Sheets."

11. Tiara Walters. "'Gentleman's Agreement': Despite Mining Ban, Russia Scours Antarctica for Massive Fossil Fuel Deposits." *PBS*, 23 May 2022, pbs.org. Accessed 5 Mar. 2024.

12. "National Petroleum Reserve in Alaska." *Bureau of Land Management*, n.d., blm.gov. Accessed 5 Mar. 2024.

CHAPTER 8. GETTING INVOLVED IN LAND PRESERVATION

1. "Urban Forests." *US Department of Agriculture Forest Service*, n.d., fs.usda.gov. Accessed 5 Mar. 2024.

2. Fiona MacDonald. "Watch David Attenborough's Emotional Last Plea in the *Planet Earth II* Finale." *Science Alert*, 13 Dec. 2016, sciencealert.com. Accessed 5 Mar. 2024.

3. "Why We're Committing to 30x30." *Nature Conservancy*, n.d., nature.org. Accessed 5 Mar. 2024.

4. Joe Ragusa. "Leslie County Teen Becomes Youngest Known Kentuckian Elected to Office." *Kentucky Teacher*, 30 Nov. 2022, kentuckyteacher.org. Accessed 5 Mar. 2024.

5. "Ministry for Foreign Affairs Invites Climate Activist Helena Gualinga as Keynote Speaker for the World Village Festival." *Ministry for Foreign Affairs of Finland*, 24 May 2023, um.fi. Accessed 5 Mar. 2024.

6. Emma VandenEinde. "Indigenous Youth Find a Way to Preserve Ancestral Lands through Conservation Group." *KUNC Radio*, 17 Aug. 2023, kunc.org. Accessed 5 Mar. 2024.

7. "Why Protecting 30% of Lands and Waters Is Critical." *Wilderness Society*, 5 Mar. 2021, wilderness.org. Accessed 5 Mar. 2024.

INDEX

ABOUT THE AUTHOR

A. W. BUCKEY

A. W. Buckey lives in Brooklyn, New York.